LET GEOGRAPHY DIE

LET GEOGRAPHY DIE

CHASING DERWENT'S GHOST AT HARVARD

ALISON MOUNTZ AND KIRA WILLIAMS

The MIT Press
Cambridge, Massachusetts
London, England

The MIT Press
Massachusetts Institute of Technology
77 Massachusetts Avenue, Cambridge, MA 02139
mitpress.mit.edu

The MIT Press would like to thank the anonymous peer reviewers who provided comments on drafts of this book. The generous work of academic experts is essential for establishing the authority and quality of our publications. We acknowledge with gratitude the contributions of these otherwise uncredited readers.

This book was set in Bembo Book MT Pro by Westchester Publishing Services. Printed and bound in the United States of America.

Library of Congress Cataloging-in-Publication Data

Names: Mountz, Alison, author. | Williams, Kira, 1990– author.
Title: Let geography die : chasing Derwent's ghost at Harvard / Alison Mountz and Kira Williams.
Description: Cambridge, Massachusetts : The MIT Press, [2025] | Includes bibliographical references and index.
Identifiers: LCCN 2024046963 (print) | LCCN 2024046964 (ebook) | ISBN 9780262551595 (paperback) | ISBN 9780262381949 (pdf) | ISBN 9780262381956 (epub)
Subjects: LCSH: Geography—Study and teaching (Higher)—Massachusetts—Cambridge—History—20th century. | Whittlesey, Derwent Stainthorpe, 1890–1956. | Geographers—United States. | Gay college teachers—United States. | Homophobia in higher education—United States. | Harvard University.
Classification: LCC G77.H37 M68 2025 (print) | LCC G77.H37 (ebook) | DDC 910.92 [B]—dc23/eng/20250310
LC record available at https://lccn.loc.gov/2024046963
LC ebook record available at https://lccn.loc.gov/2024046964

10 9 8 7 6 5 4 3 2 1

EU Authorised Representative: Easy Access System Europe, Mustamäe tee 50, 10621 Tallinn, Estonia | Email: gpsr.requests@easproject.com

To Derwent and to everyone who hid their love from view for fear of the punishments, erasures, and violence that could ensue, and to the activists who continue the fight today for LGBTQIA+ lives, rights, and liberations.

Continuities

Nothing is ever really lost, or can be lost,
No birth, identity, form—no object of the world.
Nor life, nor force, nor any visible thing;
Appearance must not foil, nor shifted sphere confuse thy brain.
Ample are time and space—ample the fields of Nature.
The body, sluggish, aged, cold—the embers left from earlier fires,
The light in the eye grown dim, shall duly flame again;
The sun now low in the west rises for mornings and for noons continual;
To frozen clods ever the spring's invisible law returns,
With grass and flowers and summer fruits and corn.

WALT WHITMAN

CONTENTS

ACKNOWLEDGMENTS

Many, many geographers and other scholars corresponded and talked with us, read chapters, and supported important conversations and insights about this writing and their own over the several years since we began this work. They include Luke Ashworth, Ishan Ashutosh, Trevor Barnes, Richard Wright, Mona Domosh, Tim Cresswell, Win Curran, Lisa Bhungalia, Jack Gieseking, Rachel Silvey, Alison Crosby, Natalie Oswin, Margaret Walton-Roberts, Kim Rygiel, Jennifer Hyndman, Beverley Mullings, Kate Motluk, Ana Visan, Corey Johnston, Sue Bunce, Matt Farish, Imre Szeman, Elspeth Brown, Beverley Mullings, Emily Gilbert, Momin Rahman, Sean Lockwood, William Walters, Kate Coddington, Jenna Loyd, and Phil Steinberg.

Tara Vinodrai, Shiva Mohan, Helen Watkins, and Becky Vogan kindly and astutely read and commented on specific chapters, and we are grateful for their thoughtful feedback.

Many thanks to Rowan Flad and Lisa Bhungalia for making trips to take photographs of key locations in this book that we could not get to in time, on the Harvard campus and in Wooster, Ohio. Thanks to cartographer Trina King for the maps in this book and Riya Osti for the index.

At Harvard, we thank our colleagues at the Canada Program and the Weatherhead Center for International Affairs and are especially grateful for the support of Helen Clayton at the Canada Program. We also learned from conversations along the way with Peter Bol, Rowan Flad, Elaine Bernard, John Augelli, Matt Wilson, and Olav Slaymaker about Harvard.

The Canada Program at Havard's Weatherhead Center for International Affairs hosted Alison as William Lyon Mackenzie King Visiting Professor of Canadian Studies, and research funds from Alison's Canada Research Chair in Global Migration at Laurier funded Kira's time at Harvard.

At Wilfrid Laurier University, we thank colleagues at the Balsillie School of International Affairs, the International Migration Research Centre, and the Department of Geography and Environmental Studies.

We are equally grateful for support from new colleagues at the University of Toronto Scarborough and the Haven Lab. Many new colleagues and staff at UTSC have welcomed and supported us from various locations including the Office of the Vice-Principal of Research and Innovation (OVPRI), the Office of the Vice-Principal and Dean, the Principal's Office, and the phenomenal librarians. We are especially grateful to Mark Schmuckler, Kristine Peruzzi, and the leadership and staff of the OVPRI and the Department of Human Geography.

Archivists in several locations supported this research. Sincere thanks to the work of librarians at the archives of Harvard University, Johns Hopkins University, Wooster College, University of Wyoming, and The History Project in Boston.

We are grateful to the MIT Press, its staff, and especially editor Katie Helke, who believed in and supported this project from the moment it arrived in her inbox, to editor Justin Kehoe, who continued this commitment and support with thoughtful advice and patience.

We also thank anonymous reviewers whose provocative questions and suggestions pushed us in just the right directions.

Portions of chapter 4 were published in 2023 by the *Annals of the American Association of Geographers* as "Let Geography Die: The rise, fall, and 'unfinished business' of geography at Harvard," reprinted by permission of the publisher, Taylor & Francis Ltd (http://www.tandfonline.com). Portions of chapter 8 are reprinted from a chapter we published in a collection, *A Place More Void,* edited by Paul Kingsbury and Anna Secor. Again, we are grateful for permission to reprint and to the editorial support along the way to those earlier publications.

We thank our friends, family, students, and colleagues for believing in us and in the importance of this story; our queer communities and friends; and Tara, whose love and patience supported Alison and this work. Finally, we are grateful to one another for our shared work, for the deep friendship, and shared personal and professional lives that sustained us and this rewarding creative work over many years on the path to its completion.

Imagine a room with tall, arched windows, late-morning light streaming in from two sides and falling on wooden floors, worn in that softened, polished way that the years have on dark wood as old as the building itself. You can smell the antiquities in the air of this room, sense the weight of centuries of debate and discussion. Forebearers in gilded frames look down as those seated around the room make decisions that will influence who and what is taught on this campus for generations to come. Here, in the Faculty Room, University Hall, you can feel the tension among powerful men— opinionated men. Harvard's President James B. Conant remains decisive, principled to the point of being stubborn, unwavering, willing to be in the minority, to step on some toes. Harvard alumnus, Johns Hopkins president, and new member of the board of overseers, Isaiah Bowman, disagrees with Conant but not forcefully, not publicly, not yet. The time is 11:00 a.m., and the board members sit on three sides of a large square table. As Bowman describes the scene, "In all, 23 men were in the room."[1]

Today is Monday, the eleventh of October, 1948, and Harvard's Board of Overseers is meeting to discuss the fate of geography as a field of study at the college. Although geography has existed at Harvard since the seventeenth century and has recently surged in popularity with the war work done by virtually every geography student and faculty member, its future remains uncertain. Just ten months ago, on January 13, President Conant initiated a policy to close the program. Although the president made this decision unilaterally, as he is wont to do to the dismay of the powerful board members around the table, the board must now review this decision. Board members have prepared a report for the meeting, authored by members Lawrence Coolidge and Charles Wyzanski Jr. Both men belong to Boston's aristocracy, to which President Conant is fundamentally opposed (Hershberg 1993, 76). In fact, Conant has come into regular conflict with the board throughout his term as president, which began in the 1930s, due to his tendency to operate without their consent. In the debate about whether geography will survive as a field of research and instruction at the college, this conflict reappears and will shape events to come.

Because of the public outcry on campus that followed President Conant's decision, the board has decided independently to review geography's situation (*Harvard Crimson*, March 4, 1948). With the board's decision, the administration has assigned its newest overseer, Isaiah Bowman, the task of assessing the members' report (Morris 1962). Although he was born in Canada, Bowman is a leading American geographer and geopolitician. At stake are three faculty positions held by assistant professors and lecturers up for tenure and promotion in Harvard's Geography Program. Bowman also chairs the external review committee of the most prominent, and arguably the most important, case among these: the tenure and promotion of assistant professor of geography Edward Ackerman.[2] According to Harvard's "up-or-out" rule, Ackerman and his faculty colleagues will be terminated if they are not tenured and promoted. Although three committees have approved of Ackerman's tenure and promotion, Conant has asked Bowman to review the situation.

The jobs and livelihoods of three Harvard geographers are on the line here, yes. But so, too, is the future of geography itself. Bowman knows this, and so does Conant.

The history that we recount in this book is a story that has been told for decades, passed along through generations of geographers, who often hear it first as graduate students in introductory courses. But this story is full of secrets and mysteries that have been hidden for years, with truths that died along with the main secret-keepers: a handful of people who were central to Harvard's decisions at the time. The story has been passed in partial form and committed to memories, we believe, incorrectly.

Until recently, only two written accounts of this history existed, the most well-known a short, speculative, yet widely read and oft-cited essay by the late geographer Neil Smith (1987), based on his extensive research on Isaiah Bowman. Before that essay, in the 1960s, a Harvard doctoral student in education named Rita Morris wrote a lesser-known but more extensive version. Other geographers have written brief, also speculative accounts, speculative even though, like the late geographer Saul Cohen, they were students on campus at the time (Cohen 1988; Martin 1988). However, the fullest information was released only recently in archival form. These university archives inform our analysis. Scattered across faculty and administrative archives, stored deep in cool underground storage, the truth has lurked all these years beneath the very grounds on which our main protagonists walked, lived, breathed, and battled with each other. As such territorial struggles tend to happen on university campuses, the fate of geography at Harvard unfolded in muted fashion, hidden behind closed doors.

As we will explain, Neil Smith (1987) infamously made Bowman the villain of geography's end in 1948, based on the power of Bowman's position, the authority he was misperceived by Smith and others to have exercised in this situation, and his secret antagonisms to parts of geography at Harvard. Given that Smith's was the dominant account, this notion of Bowman as villain has persisted in the memories and shared oral histories of geographers.

But Smith's analysis of Bowman's actions cannot be the final word. Long after the loss of Professor Derwent S. Whittlesey, geography's last permanent staff member, in November of 1956, a mythology emerged about the field's demise, tying together queerness, McCarthyism, and the politics of scientific legitimacy. With access to new documents, unavailable to either Neil Smith or Rita Morris, we have been able to read, analyze, and assemble an updated narrative, more deeply informed by empirical truths. These truths have been hidden in the archives of James B. Conant, Isaiah Bowman, and— most importantly—Derwent Whittlesey, a political geographer recruited from the University of Chicago in 1928 to build the Human Geography Program at Harvard. Although central to what happened, Whittlesey's role in the demise of geography has been misunderstood and misrepresented, his perspective missing all these decades. The fulsome personal and professional record of correspondence in his archive allows us to tell this story more fully, revealing Whittlesey's humanity, professionalism, and voice.

The story that geographers remember goes something like this: Harvard chose to shut down geography because it was run by a group of gay communists, still occasionally referenced affectionately by some people on campus as "gay commies." Because this story had wider repercussions in American geography as a whole, this mythology became legend—a shared belief about the discipline's history and source of much speculation for geographers. By choosing to "let the department die," as Conant wrote,[3] the university raised a specter that continues to haunt the discipline and the campus to the present day.

But what *did* happen to geography at Harvard during the fateful events of 1948? And what role did the death of this geography program play nationally as other universities shut down programs in the years that followed? We arrived on campus as two queer political geographers in the 2000s.[4] Rumors of what had happened to geography were unavoidable. They seemed to pursue us, unsolicited, presenting themselves at unexpected moments. Curious about the gaps in Smith's account and wanting to shed new light on his work, we reread and discussed his essay and then, in 2015, began detailed research into the history of geography at Harvard. This research program eventually led us to multiple archives in the United States covering various themes, cities, and people, including Whittlesey, Conant, Bowman, Ackerman, and Whittlesey's close friend Emeline McSweeney.

As our work progressed, we became increasingly haunted by the silences and absences that surrounded us, discovering a story of love and resistance by Whittlesey and his life partner, Harold Kemp, who he had met in Chicago in 1913. The two successfully built the Geography Program from the early 1930s and shaped it into one of the country's leading human geography programs, where Whittlesey's leadership advanced the field we share: political geography.

Elements of this story remain more hidden than you might imagine. To this day, geographers still ask whether Whittlesey was actually gay and in a relationship with fellow geographer Kemp. We want to set this record "straight," which is to set it queer: Whittlesey and Kemp fell in love as young men. They moved jobs and homes, changing cities together. Government records show in the Census and in travel documents that they lived and traveled together for decades. Their closest acquaintances—friends, family, and even some colleagues—knew about their relationship and supported it. Personal correspondence in Whittlesey's files shows his love for and intimacy with Kemp, a story we tell in more detail in the chapters that follow. Kemp and Whittlesey were a couple for over four decades.

Whittlesey and Kemp also shared their daily family life with a third geography colleague, Edward Ackerman. As we uncovered deeper links between our and Whittlesey's personal and professional lives, evidence began to reveal both institutional and individual homophobia against Whittlesey, Kemp, and their close friend and colleague Ackerman, who lived with them, traveled with them, and clearly loved them. Further, because the men were queer and because human geographers were assumed by others to be socialist at the time, Conant and conservative politicians led by Joseph McCarthy suspected that Harvard's Geography Program was Communist (Diamond 1992). President Conant associated social science fields, including human geography, with Marxism and argued that they lacked scientific merit (Conant 1948, 199). In fact, Conant used this argument to develop his 1948 policy. "I have never been able to see that there was a real need" for the program, he stated, because geography wasn't "a science in any proper use of the word."[5]

These links and this evidence showed that the legend of the end of geography in 1948 had more substance than we, Smith, or Morris knew. Given the compelling nature of the story we uncovered, we wrote this book to breathe life into the intertwined histories of Derwent Whittlesey and geography at Harvard and to restore Whittlesey's humanity to these accounts. We begin by looking back at two entangled deaths: the Geography Program's in 1948 and Whittlesey's eight years later in 1956. To recount this history, we engage not only with existing accounts but with research on the sociology of haunting, which enables us to visit the oppressive specters of gender, masculinity, and homophobia that haunt the archives. Along the way, the archives become their own set of living, breathing characters in this story. There, we stage conversations with ghosts. And in this book, the ghosts of the archives have the final say. Let us begin the chase after Derwent's ghost.

INTRODUCTION: LOVE AND LOSS IN THE ARCHIVES

THE STORY

The first time that Alison was a visiting faculty member at Harvard's Canada Program, a geographer housed in the Department of Government during academic year 2009 to 2010, a colleague with a long institutional history as alumnus and employee stopped by her office on the second floor of the Weatherhead Center for International Affairs. Standing with his body half in and half out of her doorway, he asked if she knew the history of what had happened to the geographers—her predecessors—on campus. He mentioned that they were shut down because they were "gay commies," and then he disappeared down the hallway. In fact, on several occasions during two separate years Alison spent as visiting faculty on campus, Harvard faculty members and students asked her whether she knew anything more about her "gay, commie" predecessors. Contemporary Harvard students, like *their* predecessors, were hungry for geographic education—for both insights from the discipline and explanations for its absence. The void clearly troubled not only geographers but also the Harvard community. These conversations emerged with some regularity, fueling our curiosity.

While working at Harvard as visiting faculty and doctoral fellow, we frequently encountered the mythology of the college's deceased Geography Program. After hearing rumors and speculation, we decided to seek out different versions of this story from different people, including faculty members, committees, and Harvard administration. We began by researching queerness and the discipline of geography in the Boston area from the 1920s to the 1950s. As we immersed ourselves in geography's (after)lives at Harvard, we explored the presences and absences in this mythology, discovering how the discipline's life and institutional history became inextricably tied to erased groups, such as women, people of color, and queer folk.

Our most important archival sources at Harvard were those of Professor Whittlesey and President Conant. Each left behind thick folders dedicated to the disappearance of geography. While Conant's presidential archives were more curated and orderly, Whittlesey's archives were remarkably complete in a different way: full of correspondence that was both professional and highly personal and, at times, disorganized. The files told detailed stories of his personal and professional lives, with love letters, personal correspondence, and physicians' notes. Several months into our research, we discovered that we had each developed a different assumption as to why Whittlesey's paper records were preserved so fully. We had learned early on that Whittlesey died suddenly of coronary thrombosis, mere months before his expected retirement. One of us had assumed that Whittlesey's survivors wanted his sexual relationships and life to be known in the future, thereby submitting his archives in full. The other of us assumed that his survivors were so shocked by his death that they had simply handed over his archives without reviewing their contents. Eventually, years into our research, we arrived at a third, more likely possibility: perhaps Whittlesey's partner and friend were not allowed access to his files after his death. Because Harvard had terminated their employment years prior and had never recognized their controversial intimate relationships, the university may have denied them access to Whittlesey's files, which were likely still in his work office.[1]

We also consulted former university President Isaiah Bowman's archives at Johns Hopkins University. Bowman was geographer, geopolitician, Harvard alumnus, and member of Harvard's Board of Overseers at the time of this history. As we shall explain in more detail, Conant asked Bowman to chair the external committee tasked with reviewing Ackerman's promotion file while Bowman was on the board. In contrast with Whittlesey's archives, which we believe were entered without preparation or review, Bowman spent years preparing his own archives for future generations. They included memos directed to "future men of some importance" who were over the age of forty, both requirements he made of people wishing to access the archives after his death. (Incidentally, we could not have met these conditions.)

Throughout this book, the archives help us to address unfounded empirical claims that compelled Harvard administrators and faculty members, and, therefore, us—as researchers—to examine the closure of geography. Our investigation into the two entangled deaths of Whittlesey and geography unfolds somewhat chronologically in the chapters ahead, our effort to lay out a "straight"—if inextricably queer—retelling, driven by new empirical evidence. Because of Harvard's rule that faculty and presidential archives be sealed for fifty and seventy years, respectively, after death, the university released this material only recently. In the chapters that follow, we will return all the way to the beginning of geographical instruction at Harvard in 1642 and carry this historical record to the present to trace the (after)lives of geography. Our analysis uncovers the roles of gender, masculinity, homophobia, territorial institutional politics, and secrets as geography rapidly expanded but was then starved of resources. We find, particularly, that Harvard University, under President Conant's leadership, initiated an explicit policy to "let the department die," despite the protests of Whittlesey, Bowman, and many other faculty, students, university committees, and alumni. Our analysis not only offers a deeper understanding of why the program ended at Harvard but also explores why it failed to return later and how this absence continues to haunt both the discipline and the university to this day.

We argue that this closure was fueled by the feminization of both a body of knowledge (geography) and the queer bodies of geographers involved in making geographical knowledge. We show that the same love that propelled Derwent Whittlesey and Harold Kemp forward in their personal and professional lives for decades also fed the homophobic destruction of the work that they moved to Cambridge to achieve. After Harvard administrators recruited Whittlesey to build a program in human geography, they sent him a prescient offer letter that invited him to prove that he was "the right man for the job." Sadly, we now know that Whittlesey, as a queer man, would never perform the right masculinity for the position.

By engaging with ideas about haunting, we also suggest that this story lives on for geographers in the present. Appropriately, then, this story is not strictly linear. Instead, our narrative moves forward through time, only to lurch—occasionally—into the present and then backward again to consider another person's perspective on the same events. This approach mimics the nature of haunting itself, wherein past traumas, unresolved, find expression in present oppressions, fragmenting the linearity of time. Given the importance of social location, we focus on different positionalities in the form of selected key characters—each embodying their own unique perspective in the events recounted. In chapter 4, we also juxtapose this "queering" synthetic angle with a straighter institutional lens.

THE HAUNTED PRESENCE OF TWO GEOGRAPHERS AT HARVARD

The more we read in Derwent's archives of his professional and personal life, the more compelled we felt as queer political geographers to restore life to Whittlesey's memory. Too often, written accounts describe Whittlesey as too "weak" for the job. Having spent years with Whittlesey's own files and volumes of correspondence with and about him, we find this descriptor of weakness to be a homophobic characterization of an extraordinarily successful, hard-working academic. Whittlesey was a leader in his discipline, not only at Harvard but nationally as a scholar of political geography, president of the Association of American Geographers, and long-time editor of the AAG's flagship journal, *The Annals of the Association of American Geographers*. We felt haunted by Derwent's ghost to tell *his* personal story in tandem with geography's, rather than simply repeating public records of his many career achievements. Although he has been portrayed as ineffective during negotiations to shut down the program and that mythology reinforced this image, Whittlesey was far from "weak" in character. On the contrary, as his archives show, Whittlesey was remarkably strong, bright, and beloved as colleague, instructor, and advisor. We seek, therefore, to illuminate geography's silenced history at Harvard while returning this history of geography to human form as queer embodiment.

This history haunts contemporary geographers well beyond the Ivy League, where most of them now teach. After Harvard closed its Department of Geography, other prominent universities followed suit. The only exception to this trend at Ivy League schools has been the thriving Geography Department at Dartmouth College—the place where Alison discovered the field as an undergraduate student only by stumbling into a course on economic geography taught by Professor Richard Wright. Most US-based students discover and learn the discipline in a similar way, if they are lucky enough to attend a university with faculty members who offer the subject.

THE LOFT AND THE ARCHIVE AS CLOSET

The history of geography at Harvard University remains "unfinished business," to quote the university report written by a special committee formed to examine the issue in the 1950s (Morris 1962, 3). While the story about

Figure 1.1
Photograph of Professor Derwent Whittlesey, February 1944. Courtesy of Harvard University Archives.

what happened to the discipline, its professors, and its practitioners has attained mythological status among geographers and Harvard staff and faculty, it is rife with unfounded claims and gaps in data.

As an example of this mythological status, consider the following excerpt of a long interview with University of British Columbia geography Professor Emeritus Olav Slaymaker, circulated on the national LISTSERV of the Canadian Association of Geographers. Slaymaker is a physical geographer who did his master's degree at Harvard in the 1950s, arriving on campus from London to study geography only to learn that the program had just ended. While Slaymaker later explained that he knew very little of what happened, he nonetheless felt confident sharing the following with interviewer Loretta King:

> L: William Morris Davis had been at Harvard and he had founded the American Association of Geographers, and he had a very strong legacy at Harvard, I would assume, so I'm actually surprised to hear that they had dissolved the Geography Department in 1954. Was that a reaction to William Morris Davis or do you know?
>
> O: I hate to say it. There was a scandal. Homosexuality had gone rampant. Coming from King's [College] this didn't surprise me.
>
> L: In the Geography Department?
>
> O: In the Geography Department. There was a guy called Erwin Raisz who was a cartographer, who was still hanging around the department when I got there, and he told me all the sordid details. Okay, this is not official. It's not in the literature officially. It's recorded in the literature as being the ambition of the Dean at that time to actually make a cut in the program, and this was the opportunity for him to do it because of this scandal. And I think that was the motivation of the Dean, but the substantive reason in the context was that the department was being scandalous in society. Even Harvard was very slow on new sexual ethics.
>
> (SLAYMAKER 2020, 8).

Our own entrance into this story begins with an even more casual conversation between two queer political geographers in Mountz's office in the Canada Program at Harvard's Weatherhead Center for International Affairs, located just two blocks north of a place known affectionately among local geographers of the 1930s and 1940s as "The Loft." The Loft was, in fact, two apartments occupied jointly by three geographers on faculty at Harvard at the time: Derwent Whittlesey, Harold Kemp, and Edward Ackerman. Alison suggested to Kira that we reread and revisit this history while we were on campus, where its events took place. After the reread, we decided that revisiting the archives would be worthwhile, as Smith's essay raised many questions. As it turned out, the archives held greater empirical detail on the events surrounding the program than we could have imagined at the time of these early conversations.

Although Whittlesey and Kemp were a couple for over forty years, contemporary geographers frequently ask us, "Was Whittlesey actually gay?"

To us, this question reveals the casual homophobia shrouding this history. We cannot blame geographers for not knowing this identity. How could they? It had to remain hidden at the time.

Ackerman, known affectionately as "Ed," was one of Whittlesey and Kemp's closest friends, a former-student-turned-colleague, a roommate, and the executor of their wills (Kemp 1960; Whittlesey 1956). The precise nature of their relationship is one of the more controversial aspects of this story, and we, therefore, address the relationship carefully (like all others in this book), based on facts and conversations in the archives. What we know with certainty is that Edward met Derwent and Harold as their student and eventually lived at The Loft with them for at least ten years. Close friend and cousin Emeline McSweeney identifies Ackerman as a member of their family, and their correspondence demonstrates that they clearly love each other. Ackerman is also at the heart of the institutional history we tell here, as the eventual denial of his tenure ended geography at Harvard.

Through evidence in the archives and in published obituaries, we came to know The Loft as a place memorializing these scholars and their friend-ships with students. But we also came to know it through the stationery printed by its occupants that read simply "The Loft, 20A Prescott Street." This single name, used for two separate units, indicated that the apartments together constituted one home. This arrangement, at once physical and semantic, allowed for the partial protection of a loving but illicit same-sex relationship hidden in plain sight at the heart of the institution that had developed a "Secret Court" of university administrators assembled in the 1920s to purge the campus of homosexuality (Wright 2006).

Whittlesey, Kemp, and Ackerman lived together during a time when homosexuality not only was illegal but also would cost a person their live-lihood and social stature. During its existence, the Secret Court routinely expelled Harvard students and resulted in the firing of staff, some of whom went on to commit suicide (Wright 2006). At this time, purges of queer teachers were underway in the United States, as exemplified in the more general Lavender Scare and the more specific Johns Committee in Florida (Braukman 2012; Graves 2009; Johnson 2006). Meanwhile, those who suc-cessfully navigated the social hierarchy while being queer had to hide or even erase their own sexuality to survive (e.g., Shinkle 2018; Syrett 2021). And for the men who did so, even their suspected queerness threatened perceptions of their manhood at Harvard (Townsend 1996).

The relationship between Harold Kemp and Derwent Whittlesey involved years of courtship, followed by multiple personal and professional moves, before the two men could finally live together, if closeted, in Cambridge. Yet their togetherness may also have inadvertently destroyed the subject they loved, the one that brought them to Harvard in the first place: geography.

SPECTERS OF THE "UNFINISHED BUSINESS": REWRITING HISTORY, QUEERING GEOGRAPHICAL KNOWLEDGE

Two key texts present geography's history at Harvard and its main characters: a dissertation in education completed by Harvard doctoral student Rita Morris in 1962, and a short essay published in the *Annals* by geographer Neil Smith in 1987. In her dissertation, Morris (1962, 125) argues that a university-wide policy of curtailing geography began by February of 1948, when the Conant administration decided to terminate the appointments of three geography instructors. Even though she believes this moment led to the program's downfall, Morris ultimately states that crucial details remain unknown, including *who* was behind the decision. She identifies various factors as salient to the decision: the presence and financial situation of the Institute for Geographical Exploration (also called the Rice Institute after its founder, an organization we discuss in more detail later), the absence of strong leadership by Whittlesey, and a competition between geography and geology, which existed in the same department at this time.

The more widely known and cited text, which advances this history partially and somewhat incorrectly, is the essay published in the *Annals*. The late, revered Marxist geographer Neil Smith taught Alison when she was a graduate student at City University of New York (at Hunter College–CUNY, the only university in New York City with a Geography Department). Countless instructors of introductory courses in geography, including many of our friends and colleagues, use Smith's 1987 essay to teach this history. Although speculative, we believe that it was convincing because of the depth of research achieved by Smith about Isaiah Bowman.

In the essay, Smith echoes Morris in advancing the idea that Whittlesey's weak character was the chief cause of the prospective department's inability to raise enough funding to survive. Smith also, however, emphasizes Bowman's alleged opposition to human geography as a discipline and thereby to

Harvard's well-established Human Geography Program overall. He argues that Bowman specifically used his position on the Board of Overseers to oppose the program and prospective department.

Our understanding of Bowman's role differs from Smith's. As we will show, neither Morris's nor Smith's argument is consistent with the archival evidence, and, therefore, each version produces, at best, a limited account of geography's decline at Harvard. Yet these two accounts advanced ideas about the history of geography that are still repeated today.

We aim to bring empirical substance missing from these accounts. Of necessity, Smith's essay was speculative because he did not have access to archival material about Harvard's more discriminatory views and actions. We partly complete Smith's 1987 story with analysis of archives housed at Johns Hopkins University, the City of Cambridge, Whittlesey's and Conant's previously sealed records, and Ackerman's records at the University of Wyoming. These documents enabled us to explore the stories of three major figures in geography's story: Derwent Whittlesey (chair of the Geography Program), Isaiah Bowman (geographer and president of Johns Hopkins University), and James B. Conant (president of Harvard University). Ultimately, we discovered that the institutional history of geography at Harvard was more complex than previously imagined and more influenced by masculinity and homophobia than previously surmised.

One of our interests lies with the territorial politics of ownership of this history by various people. Specifically, in later chapters, we analyze these histories alongside one another. Well-known geographers and historians of geography, including Smith (1987), Geoffrey Martin (1988), Saul Cohen (1988), and Ed Ackerman (1957) himself, all claim different versions of this history, placing blame on different individuals—usually either Whittlesey or Bowman—and social forces. Rather than laying blame, our focus is to more deeply understand how and why these people acted and interacted as they did while living under and being constrained by the complex structures of homophobia and university politics. We are particularly interested in the masculinities that unfold not only in the historical events but also in their recounting, as men try to lay claim to the stories of things that were done and not done by other men.

Particular forms of knowledge may lead to systemic injustices, including homophobia and sexism, based on values ascribed and power exercised in the treatment of these knowledge forms. Because belief depends on authority, general models of knowledge likewise depend on the political standing and positionality of knowers (Nelson 1992). Feminist philosophers of science have shown how dominant perspectives on the world are masculinized and presuppose white, heterosexual men as knowers (Bordo 1987; Young 1990). The dominant social situation of such masculinity and men generates and reinforces gender norms, including those found in scientific inquiry, such as positivism (Code 1991).

As we show, both the feminization of human geography as a discipline due to its construction of knowledge and subject matter and the feminization of the bodies and identities of its queer scholars, like Whittlesey and Kemp, ultimately led to the undoing of geography at Harvard. We can relate this loss to forms of what some scholars have called "epistemic injustice," wherein the ways in which we frame and produce knowledge itself can serve to reproduce forms of social marginalization (e.g., Hookaway 2010; Fricker 2007). In particular, the feminization of both human geography as a whole and the methods it used to study space and society led to the Conant administration's opposition to the discipline, while the feminization of its geographers' bodies and identities led to personal and institutional homophobic attacks. In this way, individual actions and social forces contributed to gender and sexuality becoming central to the life, death, and (after)lives of the Geography Program.

In the time since Whittlesey, Kemp, and Ackerman occupied marginalized masculinities and sexual identities on the edge of campus, the discipline has changed. We imagine that these geographers would be excited by subsequent approaches in their discipline to understand the geography of marginalized identities, bodies, and bodies of knowledge, particularly in the subdisciplines of feminist, queer, and Black, and indigenous geographies. These fields of knowledge construction address limitations, gaps, and the othering of both bodies of knowledge and bodies that produce geographical knowledge. Gillian Rose (1993) specifically argues that geographical knowledge is masculinist because it feminizes alternative knowledge forms and excludes women, relegating women and knowledge about women to the margins of geographical knowledge production. Queer, Black, and indigenous epistemologies and geographies have flourished, further complicating the history she wrote and advancing intersectional approaches to knowledge construction that include sexuality, sexual identity, racism, and racialized geographies as bodies of knowledge long excluded (e.g.,

McKittrick and Woods 2007, Larsen and Johnson 2012, Gieseking 2015, Oswin 2020, Noxolo 2022).

Whittlesey, Kemp, and Ackerman were all feminized as queer men in different ways. While they were able to secure faculty positions at Harvard through the privileges of whiteness and masculinity, they were eventually also marginalized due to sexuality and sexual identity. Despite his enormous success—as political geographer, president of the AAG, long-time editor of the *Annals,* professor recruited to build a Geography Program at Harvard—Whittlesey was also feminized as weak, depicted as under the influence of his long-term partner, Kemp (who himself was identified by one professor as a "dilettante"). Despite his professional stature, Whittlesey was characterized repeatedly and for decades as the "wrong" person to save geography (Stamp 1952). In spite of his intellectual reputation as a leading political geographer, his award-winning scholarship, and his professional leadership, Whittlesey is still remembered today—by Morris and Smith and in a more recent chapter by Wright and Koch (2009)—as "passive," not the right masculinity to handle this particular fight. We find this attitude reflected in archival materials as well. We see it in homophobic attacks by Bowman and geographer Dudley Stamp, as external assessors of the program, and in Conant's opposition to human geography as a discipline due to its feminized methodologies, which, according to him, lacked scientific rigor.

Similarly, the Rice Institute's involvement weakened geography's credentials. As later chapters detail, Whittlesey, Bowman, and other geographers saw the institute as "not serious" due to its origins in Boston's aristocratic patronage and its leadership by explorer Alexander Hamilton Rice. With concerns about geography's rigor, the university reacted strongly to this diminishing of geography as a serious pursuit. This positioning of geography as a scientifically weak discipline goes to the heart of the masculinist nature of knowledge construction itself, exposing which knowledge is accorded authority and which is othered and feminized. In the history of geography at Harvard, therefore, failed masculinities were closely associated with the failure to perform geography itself as a robust science, thereby rendering it unworthy as a subject the college could "do well," in Conant's assessment.

Such discourses and lines of inquiry also inform queer studies. Specifically, our investigation connects to queer geographies and research on the relationship between sexuality and space (Oswin 2020). It also connects to scholarship on queer people and the institutions that marginalize them (Ahmed 2021), from geographies of the closet (Brown 2005) to queer archives (Gieseking 2015), designed to recover marginalized knowledge, spaces, and subjectivities. In their association with The Loft, with queerness, and with each other, the three geographers at the heart of this story—Whittlesey, Kemp, and Ackerman—were wrongly blamed for the demise of geography.

The history of geography at Harvard also represents a haunting in the framing of Avery Gordon (2008; 2011), who argues that repression never acknowledged or contended with persists in the present. Thus, our approach to queer archives draws on a present absence. We are interested in how these geographers might have lived queer lives in a societal context that placed them at great risk of losing livelihood, family, and home if they did not keep hidden in spaces like The Loft and—eventually—the archive. As such, The Loft was both a closet and an escape where faculty and students socialized, close to campus, yet hidden. Whittlesey, Kemp, and Ackerman were assigned queerness by living lives that could never safely be named but were clearly spoken and written of, feminized, and marginalized, to the point of ending an entire educational program. The work of homophobia, while challenging to demonstrate, is a key finding of our account and intellectual contribution to this historical record.

To assemble this queer archive, we approach these geographers and their lives as fully as possible, reading the full records they left behind and rereading institutional decisions in the context of their (homophobic, heteronormative) time. With indications that Ackerman lived for a decade at The Loft, we went together to the archives of the City of Cambridge to read the death certificates of Whittlesey and Kemp, where we learned that both had named Ackerman as executor of their wills. Ever since that visit in June of 2016, we have pursued the entwined deaths of people and geography, seeking to restore queer lives to official institutional narratives, records, and histories.

The collective memory of this closure has always been wrapped in the specter of homophobia operating on university campuses and its specific, insidious incarnations at Harvard at that time (Wright 2006). Such histories are difficult to prove or document both with, and certainly without, empirical evidence.

THE HAUNTING PRESENCE OF DERWENT WHITTLESEY,
HAROLD KEMP, AND GEOGRAPHY AT HARVARD, 1929 TO 1956

Derwent Whittlesey died suddenly on November 25, 1956, at the age of sixty-seven.[2] We know that his death was sudden because we read a letter in the archives written by his physician just months earlier that same year, declaring him to be in good health. According to the obituary written by Ackerman in 1957, Whittlesey died of coronary thrombosis. Under Whittlesey's leadership, enrollments and course offerings in geography more than quadrupled during the 1930s and 1940s,[3] and Harvard affirmed Whittlesey's success by promoting him to full professor in 1943.[4] Once the Conant administration began to end the program, just as it had achieved its zenith, Whittlesey fought nearly a decade-long battle to save it. By 1956, however, he was the program's final faculty member. When Whittlesey died, the last geographical instruction at Harvard ended, never to permanently return.

As homophobia persists today in the retelling of this history, we want to dispel any question of Derwent Whittlesey and Harold Kemp's sexuality and long-term relationship. Although we made many incredible discoveries in Whittlesey's archives, perhaps the most compelling were the records of his relationship with Kemp, who helped Whittlesey to develop the Geography Program between 1933 and 1947. Broad evidence of their life together became apparent, ranging from their earliest love letters to joint travel records, shared expenses, and even their wills.

While Harold and Derwent met at the University of Chicago in 1913, by 1933, they lived at The Loft in a building now known as Greenough Hall, infamously named after Dean Greenough, who—ironically—unleashed the Secret Court to persecute queer students, staff, and faculty at Harvard in 1920 (Wright 2006). Family and friends knew of Whittlesey and Kemp's relationship, including Whittlesey's closest confidante, College of Wooster Professor Emeline McSweeney. Whittlesey's correspondence with McSweeney is so moving and important that we devote chapter 3 to these letters.

INTRODUCING KEY FIGURES

In the chapters that follow, we weave several people into the telling of this story, with our six major characters described briefly in table 1.1. Professor

Table 1.1

Dramatis personae: A brief summary of the cast of key characters

Name	Relation to Derwent Whittlesey	Brief description
Derwent Whittlesey	Self	An American queer political geographer, partner to Harold Kemp, and lead professor of Harvard University's Geography Program, from 1928 to 1956 (died 1956)
Harold Kemp	Partner	American geographer, partner of Derwent Whittlesey, instructor in Harvard's Geography Program, circa 1930 to 1948 (died 1960)
Isaiah Bowman	Professional colleague	Canadian–American geographer and geopolitician, president of Johns Hopkins University (1935 to 1948) and committee member to Harvard University Corporation's Board of Overseers (circa 1948; died 1950)
James Bryant Conant	Boss	American chemist and president of Harvard University (1933 to 1953; died 1978)
Edward Ackerman	Close friend, roommate, colleague, executor of will	American geographer, PhD candidate and then assistant professor of Harvard's Geography Program, circa 1930s to 1948 (died 1973)
Emeline McSweeney	Cousin and confidante	American scholar of foreign languages, Kemp's cousin, confidante to Whittlesey, queer woman, professor at College of Wooster, circa 1910s to 1940s (died 1958)

Emeline McSweeney is one of our favorites, the only woman who holds a key role in this history. In their frequent correspondence, McSweeney and Whittlesey were close and open with each other. Their letters combined emotional updates of what was happening in their personal lives with accounts of their struggles at work. For extended periods, both Harold and Derwent retreated to Emeline's home in Ohio when things were at their most difficult in Cambridge. While it can be challenging to locate homophobia in writing in the archives, it was on display in the letters that mapped the intimate relationship between Derwent and Harold and the struggles they faced on campus.

REFLECTIONS ON THE WORK OF HAUNTING

Another ghost haunts this book: Neil Smith. Smith's essay launched our investigation and inspired some of our previous work. As noted, his analysis suggested that Bowman was the villain of this history. Smith had good reasons to center Isaiah Bowman in the events of 1948, as Bowman had the power to deny Ackerman's promotion and influence Conant's opinions of geography at Harvard. But what exactly was Bowman's relationship to Ackerman's tenure case and Conant's policy? In our research, we uncovered correspondence on both situations between Bowman and Conant in 1947 and 1948—correspondence that Smith could not have read. As geography began to die at Harvard, Bowman regularly wrote to its program's staff and acted to influence Conant. For these reasons, we could not ignore Bowman's role, and we tell more of his story in chapter 5. But to hint at what is to come: although Isaiah Bowman was a powerful figure, we do not believe that he was the villain, or even the most important character, in this drama.

The story of what happened to Whittlesey, Kemp, and Ackerman, as well as geography at Harvard, matters to both history and the discipline. Our own proximity to The Loft represents the complexity and intimacy of the story that we tell: a story of love, intrigue, Cold War politics, secrets, and intersections between life and work, personal lives and institutional histories. Our presence at Harvard as visiting geographers prompted us to research these events, and we found ourselves investigating both the program in geography and the life of Derwent Whittlesey.

This history remains poorly understood. Geographers and members of the Harvard community pass it along through oral histories and through the main published text on the topic: geographer Neil Smith's 1987 essay in the *Annals of the Association of American Geographers*. Given how persistently the void of geography haunts the discipline and campus, the dearth of knowledge is remarkable. We aim to bolster this record for the discipline, for the university, and for histories of the wider Ivy League (e.g., Wright and Koch 2009). But we also aim to bolster it for queer archives now assembled posthumously to remember queer people like Whittlesey, who were simultaneously erased from history and blamed for society's ills.

In this chapter, we introduced figures who lived their lives in the harsh shadow of homophobic treatment. Yet far from shadowy, they were themselves agentic—powerful actors, well respected by students, alumni, and colleagues, far and wide. The chapters ahead help to restore the queer lives lived by Whittlesey, Kemp, and Ackerman. They also show how two deaths at Harvard—that of geography in 1948 and of Whittlesey in 1956—each hastened the other. For this reason, we found ourselves chasing Derwent's ghost at Harvard.

A HAUNTING: THE LIFE, DEATH, AND
(AFTER)LIFE OF DERWENT WHITTLESEY

Nothing is closer to the heart of the geographer than to show how other people view us. We are too prone to look only in the other direction.
—Derwent Whittlesey, letter to Grenville Clark, April 28, 1948

In March of 1948, Derwent Whittlesey found that he stood in the "ruins of his own world," in the words of his dear friend Emeline McSweeney.[1] For nearly twenty years, he had invested so much of his professional life in the development of geography at Harvard, fostering a growing program into a national leader in the discipline, a culmination of his life's work in the fields of human and political geography. Through collaborative efforts with his partner, Harold Kemp, geography at the college held a promising future, one with its own department, rapidly growing undergraduate and graduate student enrollments, scholarly achievements, and modern infrastructure. Despite these accomplishments, as we will explore in chapter 4, Harvard's administration began a policy to "let the department die," in turn provoking Whittlesey's personal and professional struggle to keep the program alive. These efforts would occupy Whittlesey's life until his sudden death in 1956.

In this chapter, we examine the life and history of geographer Derwent Whittlesey. Whittlesey was a queer American scholar who helped create the discipline of political geography. Born in 1890 and ultimately buried in 1956 in the small town of Pecatonica, Illinois, just 100 miles west of Chicago, Derwent moved to Chicago as an undergraduate student. There, he eventually embarked on a doctoral degree in geography at the University of Chicago in the 1910s, where he met his future partner, Harold Kemp, in the new library. Whittlesey studied under Ellen Churchill Semple, the first woman president of the American Association of Geographers,[2] and herself a student of German geographer Friedrich Ratzel, best known for his foundational ideas in political geography (Ashworth 2020; 2021).

Derwent Whittlesey was trained in the burgeoning science of *human geography*; Harvard University Dean Terris Moore hired him to instruct, build, and eventually lead a Geography Program in the Faculty of Arts and Sciences (FAS) at the college. During his twenty-eight years at Harvard, from 1928 to 1956, Whittlesey becomes a reputed scholar, beloved by colleagues and students alike, both on campus and across the country. From this institutional location and with sustained intellectual and professional leadership, he contributes significantly to the rapid growth of geography as a discipline—and to political geography as a core subdiscipline—at and beyond the college. In addition to bringing geography to its apex at Harvard, Whittlesey himself ranks among the leading political geographers of his time. Whittlesey is invited frequently to consult and lecture abroad, to visit military campuses and do what he and others refer to during World War II as "war work," in which most geographers at Harvard will eventually be engaged (see Barnes 2016).

The life history we tell in this chapter unfolds across various locations of Derwent Whittlesey's life, including Pecatonica and Chicago, in Illinois, and Cambridge, Massachusetts, from 1890 until his death. Correspondence in his archives shows that Whittlesey travels frequently, studying foreign languages and conducting field research in Latin America, Europe, and Africa. In this chapter, we focus not on Whittlesey's travels but on his domestic and professional life as it takes place primarily from his home in Cambridge, shared with geographers Harold Kemp and (for ten years) Edward Ackerman. Our aim in this chapter is to begin to restore life to Derwent's ghost through the humanization of Whittlesey as we explore his story and explicitly connect his life to that of the Geography Program at Harvard.

Whittlesey remains a reputed and respected scholar. Author of award-winning works, such as *The Earth and the State* (Whittlesey 1939), his life and scholarship prove critical to the advancement of the field of human geography and establishment within the United States of political geography as one if its modern subfields, centering the study of power's construction of space and place (Painter 1995).

Whittlesey also served as editor of the "flagship," peer-reviewed journal for the foremost geographical organization in the United States for twelve years, what was then called the Association of American Geographers, becoming its president in 1944. As evidence of the success of his scholarship, Whittlesey secured tenure and promotion to associate and full professor at Harvard and won various awards during his career, including the Chicago Geographical Society's Helen Culver Gold Medal.

In addition to being an accomplished scholar, Whittlesey proves popular among students and colleagues and sought after by prospective graduate students, colleagues, and military personnel seeking his insights and expertise. Whittlesey's friends and colleagues view him as a kind, compassionate, and brilliant man, whose writing and instruction influenced the development of modern geographical and political thinking.

Whittlesey's personal life was readily and richly on display in his archives, due to the high volume of personal correspondence filed there. This includes evidence of his romantic relationship with Harold Kemp, an open secret lasting over forty years. Lovers in a dangerous time, friends, family, and even some colleagues, such as geographers Ellen Semple and W. M. Davis, support Whittlesey as a gay man living in a culture of deeply rooted personal and institutional homophobia (Johnson 2006; Wright 2006; Graves 2009). His voluminous personal and professional correspondence suggests that many other confidantes, family, friends, and some former colleagues and students also knew about Derwent's long-term partnership with Harold. As we will discuss in chapter 4, the specter of Whittlesey's love for Kemp will have critical consequences for his life and the fate of geography at Harvard, especially in the attitudes of the college's administrations with respect to this increasingly successful program.

So why does Derwent stand in the ruins? Why do these ruins surround his particular body, destructive forces isolating him in the midst of so much professional success? The feminization of Whittlesey's person and scholarship particularly fosters an environment where Harvard President James B. Conant challenges both his right to head geography and the field's scientific legitimacy (Mountz and Williams 2023). As we will demonstrate in chapter 4, the feminization of an entire field of study extended not only to a body of knowledge (geographical research) but to the very bodies of those who conducted geographical research. This challenge affected the program's other faculty members, most notably Dr. Edward Ackerman (whose role we will discuss in more depth in subsequent chapters), compromising their ability to succeed and thrive as geographers on faculty at Harvard.

But who was Whittlesey, and what was his story? As two queer political geographers working at Harvard ourselves, what began as a dive into institutional archives surrounding the college's Geography Program quickly became a haunting experience and an obsession as we recognized how our own professional lives ran parallel to Whittlesey's. Although central to our field's foundation and geography's life and death at Harvard, history had largely erased Whittlesey's and geography's legacies on campus and in the wider field. We had taught, researched, published, and edited work in political geography for twenty years, yet knew little about the scholar or his scholarship. For us, what we had begun as a project to understand geography's and Whittlesey's seeming absence in the present quickly became explained by both Whittlesey's personal history and the institutional politics surrounding the end of geography. It turned out that telling the Geography Program's story without Whittlesey's was not only impossible but would also serve to reinforce the homophobic erasure he had suffered during his life.

PRE-HARVARD YEARS, DISCOVERING TWO ENTWINED LOVES: GEOGRAPHY AND HAROLD

Derwent Whittlesey was born on November 11, 1890, in Pecatonica, Illinois, to father Joseph Whittlesey and mother Sophia Derwent, his namesake (ancestry.com 2022). Derwent was the youngest child of four siblings: Susan Whittlesey (1875 to 1930), Walter Whittlesey (1879 to 1967) and, closest to him in his correspondence, Olive (Polly) Whittlesey (1882 to 1976) (ancestry.com 2022). According to Census records, by the time he was ten years old, Derwent lived at 1815 Elm Street in Rockford, Illinois (FamilySearch 2024). Rockford was a larger town than Pecatonica, located less than 100 miles west of Chicago, as depicted in the map of Derwent's life in figure 2.1.

Derwent likely lived in Rockford until the summer of 1909. While there, he attended Rockford High School, graduating in the class of 1908.[3] Nicknamed "Dersey" by close friends,[4] he participated in the school's extracurricular activities, notably becoming vice president of the Boys' Glee Club (ancestry.com 2022). Even at this time, Derwent's friends challenged his sexuality, with one writing to "not throw [women] overboard" and to seek a girlfriend.[5]

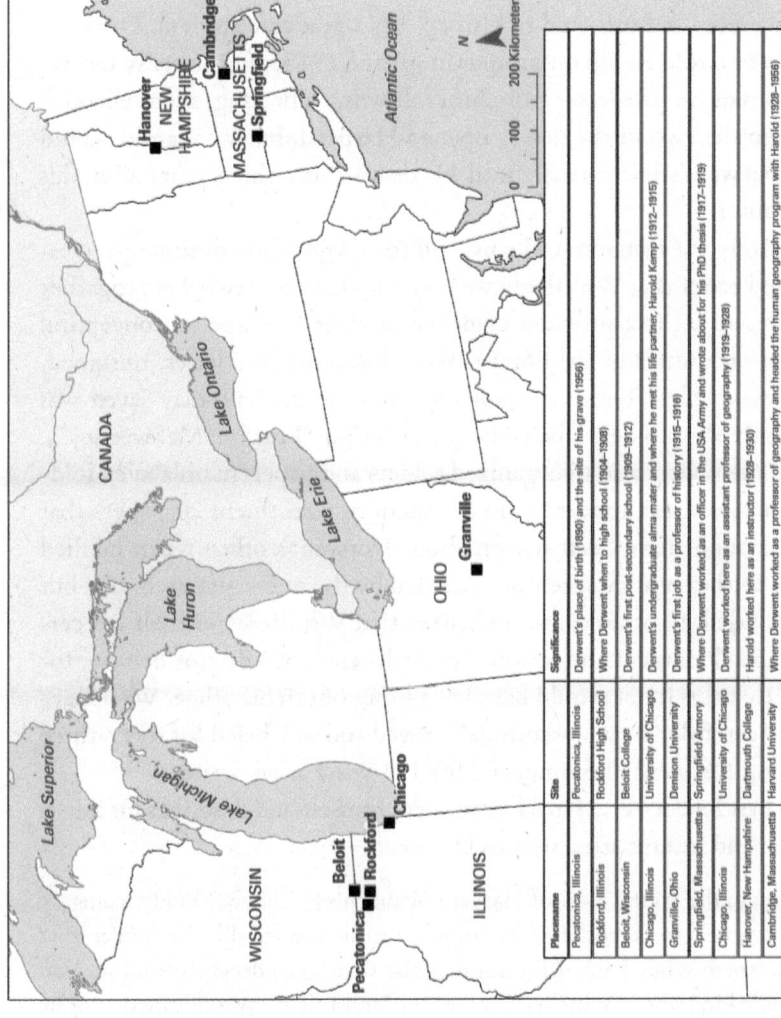

Figure 2.1

Map of Whittlesey's life showing Pecatonica, Rockford, and Chicago in Illinois; Cambridge, Massachusetts; and Wooster, Ohio. Map by Trina King.

Derwent left home by September of 1909 to attend Beloit College in Beloit, Wisconsin.[6] After spending three years at Beloit, he transferred to the University of Chicago to complete his undergraduate studies. Derwent would go on to attend the University of Chicago for graduate school, which he completed in August of 1915. As an undergraduate, he majored in social sciences and minored in history.[7] As a graduate student, Derwent would meet Harold Kemp on campus in July of 1913, after initially receiving an anonymous love letter from him following a fleeting, silent encounter between the two in the newly opened Harper Library.[8] Harold would become Derwent's life partner until his death forty-three years after this first encounter.

Fortunately for romantics like us, and for those who continue to question or need proof that Whittlesey was gay and that these two were together and in love, an artifact provides evidence of their love and its conception at this first encounter in the library. We came across the letter, unsigned, in Whittlesey's files. The correspondence that he meticulously saved was often organized in folders labeled by person (e.g., "Emeline McSweeney"). However, there were also unorganized folders and papers in unlabeled folders. Sometimes these appear to hold a random assortment of papers that we believe might have been strewn about Derwent's office when he died suddenly, neither reorganized nor discarded by the university archivists but simply included. In other cases, we believe that Whittlesey himself left certain files and documents intentionally not labeled. We do not believe, for example, that this letter would have been lying out in his office. We believe it was filed carefully away, intentionally saved and unlabeled for forty-three years during Derwent's lifetime, and for 111 years as we write.

In this first love letter, Harold introduces himself and describes his initial impression and instant attraction to Derwent:

> I saw you in the West Tower of Harper one morning, and your kindly courtesy and willingness to be bothered by stupidity made me feel all of a sudden that life was worth while [sic], if for nothing else than to be decently kind in. You put simple kindness on a higher level for me, and I thank you. Please do not be too alarmed by this. You did not speak to me nor have I ever been in the library before or since nor am I sentimental except as a human being. I do not even know your name, but if I find it out, I rather think now, that I'll mail this to you.[9]

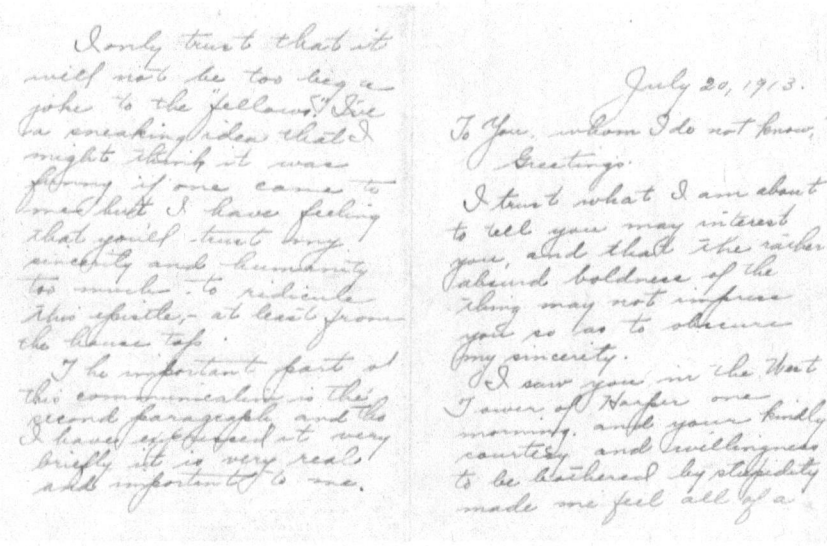

Figure 2.2
First page of Harold Kemp's first love letter to Derwent, 1913. Courtesy of Harvard University Archives, Harvard University Library.

Kemp hesitated to send the letter, however, owing to its content:

I really am taking a mean advantage but I hope you'll forgive me. Is there a law about anonymous letters? What happens? I hope you'll use it on me. But I do get so tired of seeing people and people that I like and that make me see things differently, and never making signs, that I'm just saying what I think to you—that is if I ever send it.

Three weeks later—well here goes! I'm dead sick of never doing anything that I want to and I am taking out this next sheet.[10]

In this letter, whose elaborate penmanship struck us both and inspired us to share it here, Harold uses coded humor to express his feelings toward Derwent:

It is a mean trick to work off my feelings on a perfectly unoffending stranger and down in my heart I am suffering from mortification to be doing this. Quite unnecessary you say, then why do it? Well, I've told you why. I've a perfectly good idea that you'd never do a thing on an absurd impulse, and consequently you may be very hard on a person so ridiculous. Again, I ask for mercy at your, I admit, said hands, but I think "tongue" would be more accurate.[11]

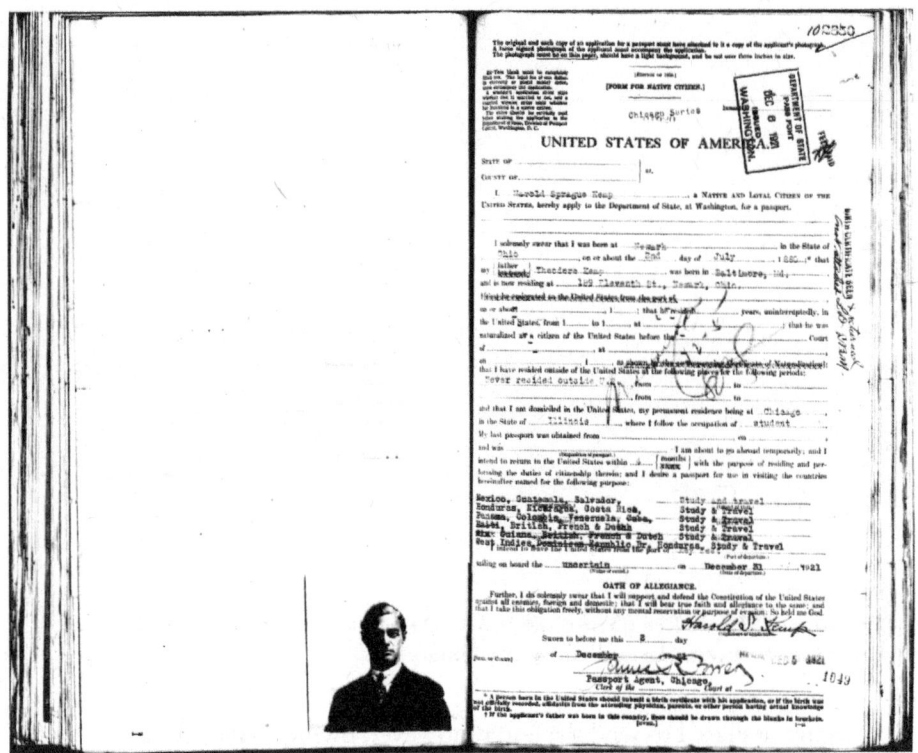

Figure 2.3
Photograph of Harold Kemp from his passport, 1921. Source: Ancestry.com.

He concludes the letter by revealing the nature of his attraction:

And now, a warning. I am not a thrilling "cute girl", so do not build a romance.
You would never imagine me guilty of this were you to see me, and furthermore you would never even see me. Besides, I am not young as you—a very
euphemistic way of stating a hard fact, *n'est-ce pas*? I have alas a strong course of
humor, which I fear that you lack.[12]

As it turned out, not only *did* Harold send the letter but Derwent
kept it for the rest of his life, preserved in the unlabeled folder in his
faculty office and then in the university archives. From this time onward,
Harold began to appear in Derwent's correspondence, especially to confidante Emeline McSweeney, which we will examine in further detail in
chapter 3.

Figure 2.4
Harper Memorial Library at the University of Chicago. Photograph by Sean Lockwood.

Harold Kemp was a striking fellow, evident in a haunting passport photograph taken of him in 1921 and depicted in figure 2.3. Surely, he was correct that he would not be mistaken for a "cute girl."

Derwent continued his education with graduate work at the University of Chicago from 1914 to 1915. During this time, he earned a master's degree in history and began work on a PhD in the social sciences. Discussing his educational trajectory in a letter, Derwent wrote: "My training was more largely in history than in geography. As an undergraduate, my major field was in Social Science [sic] with a minor in History [sic]. My graduate work was primarily in History but with a minor in geography."[13] In 1915, Denison University in Granville, Ohio, appointed Derwent as an assistant professor of history.

During this time, not only did Derwent's friends openly discuss his relationship with Harold[14] but Derwent himself noted struggles surrounding their wish to live together. Harold was working in New York City by 1916.[15] Ultimately, Derwent and Harold planned to move back to Illinois with each other, though this plan, like others, was fraught with drama and tension, to be discussed in chapter 3.

The US entry into World War I, however, changed Whittlesey's trajectory. From May of 1917 to August of 1919, he served as a second lieutenant in the US Army (Ordinance Department) in Springfield, Massachusetts, living there with Harold. This work would likely later support his activity with the Office of Strategic Services (OSS) during World War II. Whittlesey's responsibilities at the Springfield Armory included teaching an ordinance training course and acting as a private ordinance sergeant.[16] Interestingly, during this time, Harvard University offered him a fellowship, which he turned down.[17] Unfortunately, also during this time, Derwent's mother, Sophia, died on March 16, 1919 (ancestry.com 2022).

After the war, beginning in October of 1919, Whittlesey became an assistant professor of geography at the University of Chicago. His shift into geography was neither surprising nor unprecedented, as his 1920 *magna cum laude* PhD thesis, ultimately on the subject of Springfield's Armory, focused on geography as well as history, and it is very common, then and still, for scholars to move from other disciplines in the physical and social sciences into geography. Derwent and Harold lived together in Chicago, with the former being promoted to associate professor by the time he was recruited into a faculty position at Harvard in fall of 1928. By this time, the University of Chicago had become a center for the burgeoning science of human geography, and some of its key, influential scholars, such as Ellen Semple, heavily shaped Whittlesey's scholarship and professional career (Ashworth 2020; 2021).

In fact, celebrated geographer Ellen Semple served as Whittlesey's mentor. In their correspondence during the 1920s, Derwent not only sought Semple's advice on his work and career but also openly discussed his relationship with Harold.[18] Other faculty at the University of Chicago, such as Elizabeth Wallace, also knew of their romantic relationship.[19] Derwent's close friends, such as Emily Hallowell, supported his relationship with Harold, providing a place for summer retreats in Maine for the couple throughout the 1920s.

Whittlesey began his research and teaching in the relatively new subfield of political geography by December of 1922.[20] During this time, he frequently traveled for his research with Kemp, as indicated by Census and passport records, including to places such as Martinique (ancestry.com 2022).

Derwent and Harold's relationship became central to their shared future. A professional geographer who received his MS in geography in 1926,

Kemp sought an appointment as an instructor in geography at the University of Chicago but was turned down by a committee that voted 3–2 against his candidacy in June of 1926.[21] As a result, he later took up a job at Dartmouth College, in Hanover, New Hampshire, as an instructor for two years. Dartmouth placed Harold relatively close to Cambridge during the same year that Derwent relocated to Harvard: 1928.

Once again, as had been the case ten years prior, both men looked for a way to live together or nearby and, therefore, both applied to jobs in New England. Harvard University's William Davis opened a discussion with Whittlesey regarding a new Human Geography Program in 1928,[22] which offered him a professorship in April of that year—a position he accepted.

Although not without delay, Derwent and Harold's future would soon rest upon their roles at Harvard's Geography Program. Around this time, the first indications of their families' direct knowledge and support of their relationship emerged.[23] Harold's mother, in particular, thanked Derwent for his "love" toward her son.[24]

THE HARVARD YEARS: DERWENT AND HAROLD BUILD A HUMAN GEOGRAPHY PROGRAM

Derwent Whittlesey's work with Harvard's burgeoning Human Geography Program became a central element of his life from 1928 to 1956. As further explored in chapter 4, while France-based geographer Raoul Blanchard initially headed the program, FAS Dean Moore expected Derwent to share in these responsibilities and to then take over its administration within a few years.[25] In discussing his role in this prospective program, Derwent wrote to his then prospective employer, Dean Moore:

> In making my decision I need to know also how geography will be fitted into the existing organization. Shall we be a distinct department or part of some other? . . . I am particularly concerned to understand the position you contemplate for geography. Is it the intention to give it an opportunity to grow into the stature of a well rounded department, or will it be confined to a few elementary courses for an indefinite period?[26]

Although Blanchard did not formally resign from his position at the college until 1935, his frequent absences effectively left Derwent in charge of the direction of human geography, which quickly became the program's thematic focus. We disagree with the account by Geoffrey Martin that places Derwent at odds with Blanchard (Martin 1988) and instead found extensive written evidence of collegial, friendly correspondence and agreement passing between them frequently during their shared work in administering the program at this time. Correspondence between the two was warm in tone and supportive, suggesting that the two worked well together and that Blanchard supported Whittlesey's decisions and leadership, and vice versa.

During one of Blanchard's leaves in spring of 1930, Harvard University brought Kemp on as a full-time instructor for one year—a position that would be extended until 1947.[27] This new position allowed Derwent and Harold to once again live together, this time in Cambridge. Due to homophobia, however, the two lived in nominally separate though adjacent apartments at 20A Prescott Street—noted in chapter 1—which they affectionately called "The Loft." Even at this time, however, due to homophobia, Derwent hid his relationship with Harold from his family except for his sister Polly, calling himself a "bachelor."[28]

Derwent's and Harold's *working* relationship was critical to the growing success of geography at Harvard, as will be discussed in chapter 4. Despite the couple's success in being together at Harvard, Harold's employment remained precarious. It was in this context that Derwent immediately and continually fought for Harold's promotion to assistant professor for the sake of geography at Harvard *and* to be together, to the point of threatening his own resignation, as openly discussed with Dean Moore in 1931:

> Perhaps it will surprise you to learn that my chief doubt of staying here has to do with your apparent classing of Kemp. . . . His status affects me profoundly, because he is an independent co-worker, sympathetic with my own plans and competent to supplement them with his own. . . . I very much wish we might advance Kemp to the rank of Assistant Professor.[29]

No doubt, Derwent's efforts to secure a position for his life partner, a practice openly supported today by some institutions, at the time would have potentially complicated and influenced the views of members of administration, if it was known that the two were a couple.

The situation was amplified when Princeton University offered Derwent a position as professor of geography that same year. Harvard resolved this situation in 1934 only by promoting Harold to a three-year instructorship

with a much larger salary on an effectively permanent basis until his retirement. Harold's status, however, remained a topic of repeated conflict with
the new James B. Conant administration. In 1937, for instance, FAS Dean
Birkoff openly challenged Harold's credentials to be staff in the program,
which, though ultimately unsuccessful, led Derwent to contact many
prominent geographers in Harold's defense, including Isaiah Bowman.[30]
Derwent argued that "since coming here in 1930 he [Harold] has been of
inexpressible help to me. . . . If there are any brilliant geographers, then
Kemp is one of them."[31]

While seeking their own department, Derwent and Harold worked
within the confines of the Department of Geology and Geography's limitations to secure additional staff and funding, where possible. Conflict with
the college's Lowell and subsequent Conant administrations, further documented in chapters 4 and 6, emerged surrounding the creation of The
Institute for Geographical Exploration on campus, created and funded by
geographer and explorer Hamilton Rice. The Institute effectively fostered a
second site of geographical instruction at Harvard, which drew antagonism
from some leading geographers, such as Isaiah Bowman (see chapter 5), and
college officials, such as President Conant (see chapter 6). Derwent found
himself caught in the middle of these struggles, with prominent figures
like William Davis asking for his help to promote the Institute to advance
geography at Harvard.

Like Ellen Semple, Davis was aware of Derwent's relationship with Harold, to the point of naming him Derwent's "companion."[32] Derwent sum-
marizes the situation of the program in the mid-1930s as follows:

> Things [in geography] are not moving along here, partly because of the lack of
> funds and partly because it is dangerous for a new subject to expand too rapidly
> in an old institution [where] Kemp and I are trying to establish high standards.
> Even so, the registration in Geography consistently grows as does the number
> of undergraduate concentrators. . . . [Still] it has been difficult to find money
> for even our direst needs so for the next couple years we are planning to limit
> registration to the number we can take care of with our present staff. Perhaps
> that does not look like progress, but it seems to be the best solution of our
> immediate problem.[33]

Harvard's Conant administration, however, stymied these efforts in the
1930s and 1940s, refusing to allocate new funds or new staff positions.[34]

ASCENT OF THE GEOGRAPHY PROGRAM

Despite the lack of support from administration, geography as a discipline at Harvard attained its zenith under Whittlesey's and Kemp's leadership from the 1930s to 1947, growing from 102 students enrolled in seven courses taught by five personnel in 1927 to 634 students enrolled in twenty-four courses taught by twelve personnel in 1947 (Harvard University 1928; 1948). In accordance with a 1928 university memorandum, the Harvard administration eventually planned to grant the program its own department per the July 1, 1944 "Geography at Harvard" report (Department of Geology and Geography 1928; Division of Geological Sciences 1944). In this way, Derwent's original hopes for geography at Harvard when it hired him in 1928 were being realized.

Among other staff, the program hired PhD candidate Edward Ackerman as an instructor for human geography; FAS quickly appointed him as assistant professor of geography in 1939 (Harvard University 1940). Edward formed close connections with Derwent and Harold, first as Derwent's student and, soon after, living with the two at The Loft by 1940 (US Census 1940). He continued to live with Derwent and Harold until his departure, after the loss of his position in the Geography Program during the events of 1947 and 1948 (see chapter 4).

Ackerman's bond was so close yet relaxed that when, for instance, Derwent took a leave of absence from Harvard in March of 1948 to visit McSweeney, Ackerman opened and read Derwent's mail on his behalf. In an important sense, Edward's work and position within Harvard's Human Geography Program represented the growing culmination of its success and commitments, making his loss in 1948 a striking personal blow to Derwent as well as for the long-term plans for geography that he had worked to execute at the college. The fate of geographical instruction was bound not only to the embodied presence of Whittlesey but to his former student and contemporary colleague, close friend, and family member at The Loft, Ackerman.

Derwent's career and aspirations for geography at Harvard rapidly advanced throughout the 1940s. In recognition of his accomplishments, the FAS promoted him to full professor in 1943, and he was elected president of the Association of American Geographers in 1944 (Harvard University 1942; 1944). Reflecting upon his promotion, Derwent said to FAS Dean Paul Buck: "I suppose the most gratifying thing of all is your recognition that geography should have a standing at Harvard commensurate with its position in the world at large."[35] By this time, Derwent helped to found political geography as a new subfield of inquiry in human geography, with seminal works such as *The Earth and the State: A Study of Political Geography* (1939). He published his first piece on the subject in the *Journal of Education* in 1935 (Whittlesey 1935). This was also the period when Derwent began to assemble his own archival records for future researchers.[36]

DERWENT'S SCHOLARSHIP AND ROLE IN THE DEVELOPMENT OF POLITICAL GEOGRAPHY

Derwent Whittlesey's political geography was part of a burgeoning materialist tradition on the relationship between human politics and the Earth, heavily influenced by scholars at the University of Chicago—chief among them Ellen Churchill Semple. Political geographers studied, among other things, the importance of human interactions with space and the physical world through concepts such as technology, capital, land use, as well as state-building and imperial projects.

As Luke Ashworth (2021) writes, Whittlesey's research argued for an empirical, fungible, and complex relationship between the Earth and human society, specifically and explicitly rejecting the environmental determinism found in geographer and ethnologist's Friedrich Ratzel's and Ellen Semple's work. Although rejecting geography as determining human society and its conditions, he nonetheless maintained its centrality in influencing society, given that the former would always be embedded in geography itself. The culmination of this thought, *The Earth and the State* (1939), was a huge success, selected as one of the top fifty educational textbooks published in 1939, and as a text for the US Armed Forces Institute. Through various themes, Whittlesey's seminal contributions in human geography, such as those on sequent occupance, borders, and nation-states themselves, made him a key figure in the history of political geography (Ashworth 2020).

Whittlesey's research also received The Chicago Geographical Society's Helen Culver Gold Medal, one of the world's leading geography awards, in 1947 (Harvard University 1947). His work brought him into close, regular contact with other geographers working on similar research interests, particularly Isaiah Bowman, who collaborated in their attempt to distance political geography from *geopolitics*, which society had largely come to associate with Nazi German imperialism. Air University Air Command and Staff School officer Max van Rossum Daum exemplified this in a letter to Derwent in 1947, writing:

> There is another item we talked about and which is the subject of a great deal of discussion at this university, namely the word "Geopolitics". The opponents object to "geopolitics" because the originators of the term have the sole right to define it, and the German definition is largely unacceptable. . . . Since these gentlemen dislike the connotation of aggression, and since the German use ignored all ethical and moral values, they want to abolish the term and even have suggested: "Geologistics." . . . Whereas I agree the word "geopolitics" has unpleasant associations with German neoimperialism, I consider our understanding of the subject matter important information for future air commanders of the United States air forces.[37]

Rossum Daum then asked Derwent whether he was also in favor of abolishing the term, and, if so, which term he might recommend be used in its place. In his response, Whittlesey delineates "Geopolitik" as the German school, in contradistinction to geopolitics as a whole. He further proposes political geography as a "well established term, meaning the objective study of all phenomena that integrate political affairs with earth conditions."[38] He admits that geopolitics remains useful as a term but that it must be dissociated from Geopolitik, as embodied by scholars such as Friedrich Ratzel. By positing that political entities behave like living organisms and, therefore, by their very nature, must grow and expand, Ratzel's work directly contributed to Nazi German concepts of Lebensraum as both claimed scientific reality and a goal of the state (Cohen 2003). German fascists used such concepts to enact and justify their imperialist and ultimately genocidal goals.

Uncoincidentally, the strategic importance of geographic knowledge to *American* efforts in World War II precisely propelled new funding, courses, and staff for Derwent's Geography Program. The war also brought him back into service with the US military through his work for the Office for Strategic Services and as a consultant to the US Army and Navy.[39] Whittlesey, Kemp, and Ackerman also instructed US Air Force members in geography.[40] During this time, there are frequently references in correspondence to and from Derwent about the work most geography faculty and students were doing in service of the war, referred to as "war work." For Whittlesey, this involved travel to consult in Washington, DC, office training, and consultations and lectures with military personnel, as well as military service.

DERWENT'S LIFE IN THE 1940S

The 1940s eventually became a difficult decade for Derwent. On March 2, 1942, his father, Joseph Whittlesey, died (ancestry.com 2022). Despite the postwar boom in student enrollment at Harvard due to the GI Bill,[41] and even given the increased need for geographical knowledge to train postwar imperialist administrators, such as during the occupation of Japan, the college began to slowly end the Geography Program. This slow demise began with the program's struggles to secure tutorials and resources in 1945. Responding to the Conant administration's refusal to grant any new permanent staff at this time, Derwent wrote to geographer and future professor at UCLA Henry Bruman:

> We can only offer an annual appointment, and we have positively nothing in sight beyond three years, because of a rule which makes it impossible to use the same man on an annual appointment for a longer term. . . . I am sure it [geography] will expand after the war. However, any one who comes here can be given no prospect of a permanent position.[42]

In general, Derwent continued to seek a Department of Geography that would be separate from geology, based on geography's interdisciplinary nature and its need for independence. Responding to an inquiry from a Yale professor also seeking to create a Department of Geography there in 1945, for instance, he wrote:

> I have filled out your questionnaire, but it can hardly be of much value to you, because our situation here [at Harvard] is the result of a unique history, and is not duplicated elsewhere, thank God. The most important thing is to set up geography by itself, so that it can then make free and equal contacts with all border fields, from geology to political science.[43]

These sorts of struggles for geographers to explain their interdisciplinary discipline to university administrators are common and continue today. This advice, written by Whittlesey in 1945, is the same kind of advice that contemporary geographers and department chairs offer to one another in precarious times when resources are lean, budgets tightened, and restructuring looms.

At Harvard, the forthcoming conflict over the Geography Program caused the loss of most remaining human geographers on campus by 1950, except for Whittlesey himself (a process detailed in chapter 4). Whittlesey's home life, deeply entwined with Kemp and Ackerman, became complicated by President Conant's policy to "let the department die," which led Kemp into retirement in 1947 and forced Ackerman to seek employment elsewhere when Harvard effectively terminated his professorship during the 1947–1948 academic year. This policy also reflected institutional and individual homophobia, which would come out through both the administration's direct actions against queer people (as in the Secret Court) as well as coded character attacks in the archives.

A letter to colleague Charles Colby, a professor of geography based at the University of Chicago, contained what was perhaps Whittlesey's most compelling summary of geography's situation during the events of 1947 to 1948, given Conant's policy:

> It turns out that if Ackerman leaves now geography at Harvard will wind up and come to a stop. If we cannot retain him, we cannot get anybody else now whom we could get past the administration. Since Harold is retiring this year, our courses will there-fore [sic] promptly shrink to a level so low that we will have no graduate students. There is also in prospect a shrinkage of undergraduate enrollment, because of the substitution of general education for courses that students may choose, including geography. I therefore foresee only a bleak prospect for myself to continue operating without either adequate support of colleagues or students.[44]

In a haunting foreshadowing, Derwent also wrote of the wider repercussions of this policy in the same letter: "What effect it will have on other universities to learn that Harvard has abandoned geography, I hate to think. I think it may be a serious setback all along the line."[45] Hidden on the letter's backside was a note from Harold, adding, "Everything Whit Says is true. We are in a mess, just when it seemed that we had found our place in the sun."

By March of 1948, Harvard had abolished geography as an undergraduate field. In his letters with confidante Emeline McSweeney, Whittlesey noted how difficult his situation was and how unhappy he had become with it.[46] Regarding his perspective, he wrote to his colleague, geographer George Cressey at Syracuse University: "Needless to say, this whole business is a crushing blow. I have been able to do no work of any kind since it [the decision] fell and much the same is true of the rest of us. . . . To have it all knocked out from under us is hard to take."[47]

In response to the Conant administration's actions against geography, Whittlesey began a campaign to defend the program, as explored in chapter 4, reaching out to many prominent geographers and recording its accomplishments, with Ackerman writing, "I still feel that [President] Conant owes you (and me too) an explanation which has not yet been given."[48]

But no explanation satisfactory to geographers and others on campus would come. The October 1948 Board of Overseers' report reviewing the situation of geography at Harvard also alluded to this missing explanation, avoiding explicitly naming it while adding that there were "good and sufficient reasons" to end the program.[49] These hidden reasons, explored further in chapters 5 and 6, lay at the confluence of forces acting against Derwent's work and identity: homophobia, Cold War politics, and the legitimacy of geography itself as a scientific field.

Despite Kemp's retirement and Ackerman's departure for employment with the postwar Allied occupation of Japan,[50] Whittlesey continued his work, especially the defense of Harvard's Geography Program, as he described to London School of Economics geographer Dudley Stamp in August of 1949:

> We geographers at Harvard have had a bad set-back. I can only take comfort in the knowledge that geography has been assassinated many times before, but has always continued to live. I am not willing to believe that Harvard will not in time see the light.[51]

That Whittlesey would pen these lines to Stamp is significant as a precursor to what would follow, as it was Stamp who would later be contracted by Conant to assess the prospect of geography's future at Harvard.

In May of 1948, Whittlesey coordinated a joint letter protesting the Conant administration's actions with the AAG and other geographical organizations in the United States.[52] From 1948 until 1956, no fewer than five institutional reviews of the situation of geography at Harvard University occurred (discussed in chapter 6); ironically, one of these was the confidential report authored by Stamp, solicited by President Conant (Stamp 1952).

As geography's very survival flickers at Harvard, Whittlesey and Kemp maintain their relationship and joint residence at The Loft.[53] Kemp continues to support geography behind the scenes, assisting with Whittlesey's graduate students, even being affectionately called "Uncle" by some of them, like John Augelli, as he provided comments and feedback on their work.[54]

Many of Whittlesey's former and contemporary students communicate with him directly as the situation surrounding Ackerman's tenure case and the prospective closure of the Geography Program unfolded. Through these letters, we found Whittlesey to be loving and supportive and maintains close ties with these students. Some students, such as future University of Toronto professor of geography Stephen B. Jones, also personally know Ackerman and support him. In November of 1947, Jones writes: "[On the] subject of Ed's promotion. What a ghastly system of promotion Harvard has. The only thing I know about Ed that is derogatory to his intellect is the fact that he hasn't told Harvard to kiss (you know where) long ago."[55]

When the events of 1947 and 1948 occur, as we will further discuss in chapter 4, these alumni rally behind Whittlesey, with some, like Jones and Kemp's former student Peter Roll, even going so far as to contact President Conant directly to request that he reverse his policy on geography at Harvard. According to Ackerman, Roll proved instrumental in organizing these alumni "behind the scenes" in their fight to preserve the program as a whole.[56] Roll authors a letter to the editors of Harvard's student newspaper, *The Crimson*, on March 9, 1948, stating, "I can only feel that the removal of the department is contrary to Harvard's general education program . . . and to the best interests of Harvard in general. Before our valuable staff in the department takes leave of us, the edict should be carefully reconsidered" (*The Crimson* March 9, 1948). Abundant personal correspondence with students like Roll and colleagues like Jones, Cressey, and others demonstrated how close Whittlesey and Kemp remained as mentors to current and former students and colleagues, who, themselves, became invested in the program's fate.

Despite these events, Whittlesey and Kemp sustained their close relationship with Ackerman. Ackerman, for instance, wrote letters with "love" to Whittlesey,[57] and Whittlesey said of him:

> As you observe, Ackerman is not really youthful. He has never been and the combined efforts of Harold Kemp and me over a dozen years did not change him. He is, however, almost the finest person in the world to be with.[58]

These queer intimacies supported Whittlesey during his struggles with Conant's policy, and abundant discussion exists in his archives about his relationship with Kemp and the role of Ackerman in their chosen family and shared home at The Loft.

By summer 1956, Derwent was moving toward retirement and remained with Harold at The Loft. He was in good health[59] and planned to work as a sessional instructor in political science at the Massachusetts Institute of Technology.[60] Even at this time, we found evidence of Whittlesey's efforts to bring geography back to Harvard University as committees continued to examine its situation (as shown in chapter 4). Unfortunately, Derwent died suddenly on November 25, 1956 (Ackerman 1957). We discovered the circumstances and nature of his death from his academic obituary, published by Ackerman in 1957. As previously noted, Ackerman was also the executor of Whittlesey's will. In his will, Derwent bequeathed the vast majority of his estate to Kemp and Ackerman (City of Cambridge Archives 2016). Derwent's body was finally laid to rest at Twelve Mile Grove Cemetery in Pecatonica, Illinois, USA (figure 2.5; ancestry.com 2022).

CONCLUSIONS: ON LEGACIES, HISTORIES, ARCHIVES,
AND THE CLOSET

Based on records in the archives, we assembled a list of people who we believe were likely Whittlesey's graduate advisees. This list includes Jack Ransome, Saul Cohen, Mary Stewart, Stephen B. Jones, George K. Lewis, Donald Patton, Richard Logan, Rhodes Murphey, Howard L. Green, Robert Johnson, John Enman Jr., Edward Ackerman, and John Augelli (with whom we spoke only very briefly by telephone not long before he died).

Figure 2.5
Derwent Whittlesey's grave in Pecatonica, Illinois. Source: ancestry.com.

Over the course of our research, we attempted to speak with living members of this cohort and with any other people we learned about along the way who were engaged with geography at Harvard. Unfortunately, most of Derwent's student cohort had either passed away or did so during the course of our research, such as John Augelli and Saul Cohen.

Olav Slaymaker, for example, who arrived at Harvard in September of 1961, told us that not only did geography no longer exist on campus at this time but that his contacts in the former Institute for Geographical Exploration stated that the Conant administration had shut down the Geography Program due to "inappropriate behavior" tied with general homophobia. He also connected Harvard's homophobic culture with his experiences at Oxford University, including the infamous death of Alan Turing. In contrast with administrative accounts, Olav said that his contacts told him that the program's closure was ultimately more personal than external (Slaymaker 2020).[61] By this time, additionally, he noted that the discussion of geography on campus had largely disappeared. We confirmed that Olav's contacts were in fact cartographic staff in geography, where they had worked at Harvard since the 1930s.

We also reached out to family members of Derwent and Harold whom we found using ancestry.com; however, those with whom we spoke either were unaware of them or could provide no new information. Ancestral records, moreover, such as Census and travel documents and newspaper articles provided additional information about Derwent's life. Despite this, the wider absence of further sources of information on Derwent led to his archives becoming our primary source of information about his life.

Given the timing and suddenness of Derwent's death, we examined various possible accounts for why his archives were as chaotic yet complete as we found them during our research. The archives especially contained broad evidence of Derwent's relationship with Harold, including that first love letter—evidence that could have easily resulted in the loss of their livelihoods or even lives. While we initially considered that Derwent may have intentionally left such letters for the future, available evidence regarding how Harvard handled its archives instead suggested that those responsible for deciding what to do with his archives, possibly Harold or Edward, or—more likely—university archivists, may have been overwhelmed by the sheer volume of material to process, particularly during grief. It was also extremely likely that these archival records included only a portion of Derwent's correspondence, even related to his career, as implied by not only his writing in them but also in the absences of particular pieces of correspondence that otherwise should have been present. Pertinent to our next chapter, for instance, Derwent kept extensive records of Emeline's letters to him; however, he kept few copies of his own to her, even though he often made carbon copies of important, type-written letters. Furthermore, her correspondence that *is* preserved in his archives refers to additional letters not preserved in his archives. We believe that these were likely still at The Loft, and that for reasons related to privacy, Derwent kept some of Emeline's correspondence in his office at work.

We have narrated so much of Derwent Whittlesey's life history in relation to Harold Kemp because Harold was his partner for most of Derwent's adult life, from the time that they met when Derwent was just twenty-three. This is significant because of the ways that Derwent's identity and Derwent and Harold's relationship have been consistently called into question over the years, in publication (e.g., Martin 1988) and orally, with questions that geographers still casually raise. Perhaps unintentional homophobia informs these frequent questions we receive upon mentioning this project to other geographers. In general, these questions reflect a pervasive heteronormativity that, to this day, still presumes heterosexuality as the default and homosexuality as artificial, contrived, and thereby demanding evidence to support its existence. In this chapter, we have provided such evidence only secondarily to refute such claims, instead focusing on it as a celebration and rehumanization of Derwent's and Harold's lives as two queer men living in homophobic times. Next, we turn to the humanization of Derwent in the distinct and beautiful ways he comes to life through his intimate correspondence with Emeline McSweeney, also sustained over the course of his adult life.

"WITH MUCH LOVE, EMELINE": CORRESPONDENCE BETWEEN "THE REMARKABLE EMELINE MCSWEENEY" AND DERWENT WHITTLESEY

Emeline McSweeney, professor of languages, was a close friend of Derwent Whittlesey and corresponded with him devotedly for most of his adult life. Emeline lived in Wooster, Ohio, down the street from where she taught as a faculty member at the College of Wooster. She was an American scholar of foreign languages and Harold Kemp's cousin. We do not know the exact year that Emeline and Derwent met but can infer from Emeline's correspondence that their first encounter took place sometime after Derwent and Harold met in Chicago in July of 1913, and before Emeline's first recorded letter to Derwent in August of 1916. Derwent saved this letter for the rest of his life, and it was moved into Harvard's archives alongside twenty-four others after he died forty years later. It is the first in his folder, labeled "McSweeney, Emeline," a queer archive within a queer archive, a treasure buried within the treasure that is Derwent's faculty archive, depicted in figure 3.1.

Emeline's very first letter opens with the usual, direct, and disarming intimacy with which she tends to address Derwent: "Dear Derwent, Don't get alarmed and fancy I am going to answer all your letters by return mail. I simply wanted to say a few things to you before you leave on Saturday. I have just read your letter over twice and I read all that was between the lines, as well as what you said in words."[1]

This first of Emeline's letters that Derwent saved is dated August 23, 1916. We believe that the letter—which we will return to shortly—eventually made its way into Harvard's archives from Derwent's office, where it was still housed in Derwent's files when he died on November 25, 1956, precisely forty years and three months after its arrival. This letter is important to understanding the rapid level of intimacy established in the sharing of details between Emeline and Derwent. Like others we discuss in this chapter, it also provides insights into the intense rhythms and cycles of the romantic relationship between Derwent and Harold—perhaps the most frequent

Figure 3.1
Photograph of folder with the name McSweeney, Emeline handwritten, likely by
Derwent Whittlesey. Courtesy of Harvard University Archives, Harvard University
Library.

topic of conversation in Emeline and Derwent's correspondence. These relationship rhythms often involve Harold departing their shared life from time
to time for other cities, like New York and London, which leaves Derwent
fretting and corresponding with Emeline about the status of their relationship. On other occasions, the letters address Harold's attractions to and relationships with other men, such as Michael, detailed in a letter in the 1950s.

With intimacy in both content and tone, Emeline's letters establish the
close friendship that developed quickly between herself and Derwent. In
this first letter, Emeline shares her own queerness and reveals herself to be
an important confidante and witness to the trials of Derwent and Harold's
love and relationship.

The two remain faithful and fairly regular correspondents, with the
exception of a year-long period when Emeline does not hear back from
Derwent and worries that she might have upset him the last time that she

had seen him in person. Nearly three decades into their correspondence, Emeline explains, "It is always a red-letter day when a letter from you arrives."[2] Emeline and Derwent express mutual respect and admiration, shared interests and fondness for one another. This common ground fuels and sustains their friendship for four decades. In writing, they gleefully anticipate visits together, as in Emeline's letter addressed to "Der" in February of 1948: "I have been all in a glow of delight ever since your letter came, thinking that you are really going to visit me. Any time will suit me. When I heard you were coming, it gave me a glow of pleasure for several days."[3]

Even though she is often writing in response to and about Derwent's letters to her, we do not have access to any of Derwent's correspondence to Emeline. This is an anomaly within his files. During a time of carbon copies of typed correspondence, we read dozens of Derwent's letters to others, which he meticulously saved and filed. Derwent's letters were personable, warm, and thoughtful, giving us a good sense of his voice and of why colleagues and friends were so fond of him. Therefore, even though we do not have his letters to her, it is no surprise to us that Emeline looked forward to and cherished Derwent's letters, reading and rereading them, writing long, beautiful letters in response, and explaining how important his correspondence and friendship were to her.

We tried to find additional archives that would document McSweeney's work and life beyond what we found in Whittlesey's archives at Harvard. We first contacted Special Collections, which houses the faculty archives at the College of Wooster in 2016, but this initial inquiry turned up only a black-and-white photograph of Emeline, depicted in figure 3.2.

Subsequent contact with the College of Wooster archives yielded additional photographs but few additional documents. While richly portraying aspects of her personal life, the letters in Whittlesey's files say little about McSweeney's work. Nonetheless, these letters enable us to explore another perspective on Derwent Whittlesey and his personal and professional trials, and Emeline McSweeney's, as they are documented through the intimate correspondence. Emeline was aware of Derwent and Harold's romantic relationship from its early stages, offering frequent advice, refuge, and support throughout their many struggles, as they matured and moved from the University of Chicago to within commuting distance of each other in New England at Harvard and Dartmouth.[4]

Figure 3.2
Photograph of Emeline McSweeney. Courtesy of Special Collections, the College of
Wooster Libraries.

Derwent confided in Emeline aspects of his life otherwise rendered invisible to others, including his deep desire to be and live with Harold during a time when homophobia made this arrangement risky. This correspondence also documents—to some extent—the personal effects of the Conant administration's termination of geography at Harvard and Derwent's resistance to that policy.

Emeline's letters offer a different voice and perspective on queer life, institutional life, Derwent and Harold, and the events at Harvard that we narrate throughout this book through institutional politics and personal squabbles. As she ages across the decades, Emeline provides glimpses into quotidian life for queer women at that time, as she shares struggles to sustain her health and her work. Emeline reveals bits of herself in these intimate letters. She loves to read and listen to music, often alone, although she also reports on family relations and visits and women we believe to be romantic partners. She loves to sit and read alone at her window, especially in autumn: "Autumn is my season."[5]

A NOTE ON THE LETTERS, OUR APPROACH TO THEM, AND THE THEMES THEY CONTAIN

It was not uncommon for Derwent to keep folders of correspondence labeled and organized by author. In total, he had filed thirty-five letters in the folders labeled "McSweeney, Emeline." Emeline's was, therefore, among the more robust personal correspondence folders.

Emeline often wrote her letters by hand, later typing a total of six of the letters filed by Derwent. The letters themselves are beautiful artifacts, as shown in the examples in figure 3.3.

Eventually, due to deteriorating eyesight in later years, Emeline asked Edith Yoder, the woman we believe to be her partner at the time, to transcribe on her behalf. These letters were signed by their transcriber's initials "E.Y." and contain the occasional smiling face when Emeline referred directly to her scribe, Edith, in her narrative to Derwent.

Occasionally, Emeline wrote on letterhead from the college: "The College of Wooster Academy, Department of Latin and Greek." She subsequently, and most commonly, used personal stationery.[6] Eventually, Emeline used more formal personal letterhead with her home address in

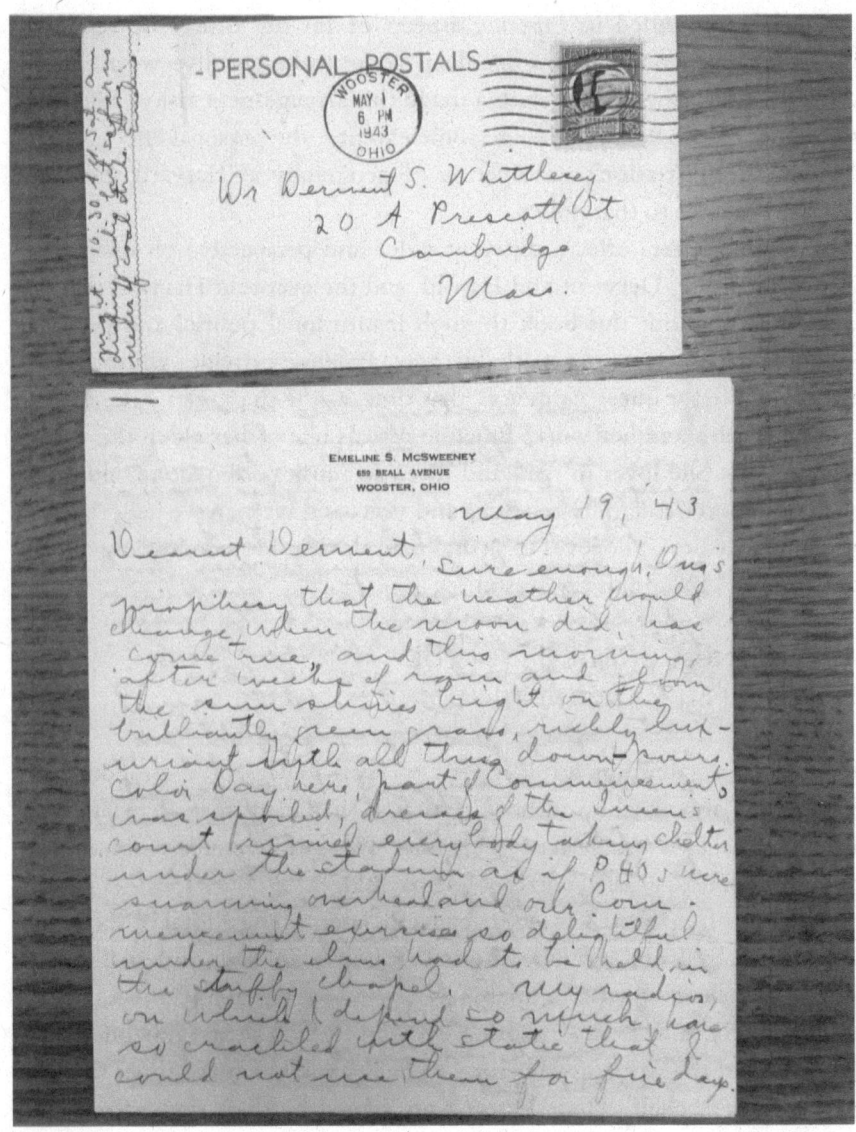

Figure 3.3
Photograph of handwritten letter and envelope addressed to Derwent Whittlesey on
Prescott Street. Courtesy of Harvard University Archives, Harvard University Library.

Wooster, Ohio, printed at the top. The first letter locating Emeline with letterhead placed her at 659 Beall Avenue in Wooster, Ohio, in a house that still stands. This letter is dated February 20, 1934. Emeline lived in this home with her mother and eventually by herself, with various male and female caregivers who come and go, and eventually with Edith Yoder in her later letters of the 1950s. The house, pictured in figure 3.4, remains on a busy street that leads directly to the college.

We know that Derwent Whittlesey's Emeline McSweeney folder does not contain all of her correspondence. Sometimes, she references letters that do not appear in this file, as in her reference in a letter dated February 17, 1948, to the personal news she shared in her January letter of that same year.[7] We imagine a few reasons the letters do not all appear in this folder. First, the archives contain the files that were moved from Derwent's campus office. Emeline sent the correspondence to his home address, and we imagine more letters were likely stored at The Loft on Prescott Street,

Figure 3.4
Photograph of McSweeney's home at 659 Beall Avenue, Wooster, Ohio. Photograph by Lisa Bhungalia.

as they all would have arrived there—most envelopes in this folder were addressed to Prescott Street, as depicted in the photograph in figure 3.3. Second, Derwent traveled frequently for work, and the two corresponded during their travels. Finally, it is also possible that Derwent did not always save each letter.

From time to time, Emeline instructed Derwent, in writing, to stop sharing or reading a letter to Harold. We interpret this as an indication that Derwent shared Emeline's letters with Harold, sometimes reading them aloud to him, but that he may have strategically brought those he chose— or she instructed him—not to share to his office, which is the file which we now, inadvertently, are able to read. We frequently found correspondence in others' archives, including the personal files of Edward Ackerman and Isaiah Bowman, with explicit instruction to keep letters and their content secret. In the case of Emeline's letters to Derwent, these instructions would be written directly into the text of the letter (an indication that a reader sharing the text aloud may wish to stop reading here).

Our approach to the thirty-five letters was to inventory and read them repeatedly as a continual text, eventually coding lines for key themes. These include friendship, work, Derwent's considerable work ethic (including "war time work," when she observes that he has two jobs during the war: one in Cambridge and the other in Washington), travel—including visits that they will make to one another—Derwent's relationship with Harold, the status of the Geography Program at Harvard, and the role that Edward Ackerman plays in Derwent and Harold's home. These themes also reflect our own developing understanding of and interest in the threads of the stories we tell in this book. Emeline and Derwent's relationship, and the revelations in their correspondence about the relationship and friendships they have with others, prompted many conversations between us about queer relationships in this time, driven as they are today by the expansiveness of queer love and platonic intimacy and by the desire to build queer families. It is fair to say that these themes emerged simultaneously from our inductive reading of the letters and based on our broader analysis of the stories. They also helped to inform how we understand and tell this history.

We structure this chapter around these themes, giving each its due attention.

EMELINE ON DERWENT, HAROLD, AND HER OWN
QUEER RELATIONSHIPS

Harold is Emeline's cousin, which is a frequent topic of conversation between letter writers. From the letters, it is clear that Harold is an
emotional character who falls hard for people, and who sometimes acts
independently of his relationship with Derwent. This causes fretting,
hand-wringing, and consternation on the part of both Emeline and Derwent. There are frequent references to not being able to manage or guide
Harold, who must, after all, find his own way, according to Emeline. While
Harold is Derwent's chronological senior, they often discuss him as though
he is an unruly youth of sorts, but their own and, therefore, one they both
love unconditionally nonetheless. Emeline writes in one letter, when Harold and Derwent's future remains unclear: "You speak of having uprooted
Harold from his whole past life and yet being unable to control his course
for the future. . . . And what a mess you'd make if you could control it,
and what a mess I'd make, much as we love him. Thank Heaven it isn't our
job!"[8] When Harold leaves Derwent to spend some months in New York,
Emeline writes: "I think you entirely overrate the loneliness of Harold's
first months in New York. You know what a remarkable capacity he has
for making friends quickly. Then, if he gets into some of his creative work,
something he can put himself into, it will absorb him enough to keep him
going straight ahead. He can throw himself absolutely, you know, into
what interests him."[9]

Like Derwent, Emeline is an academic and intellectual. She shares his
passion for reading, history, and cultural arts such as theater, music, and
photography. They often share books to read, radio programs to listen to
late at night, and stories of outings to the theater in their respective home-
towns. They loan each other books and records, sent by mail, and also
mail photographs of works of art. Alongside a postcard featuring a photograph by Breton of a woman titled *The Cleaner*, Emeline writes, "This is
the Breton I was telling you about. How do you like it?" They correspond
frequently about friendships, romantic partners, and work lives. They also
discuss their spiritual lives. Emeline entreats Derwent to visit frequently,
and they often chat about upcoming trips to see each other or about travel
to other parts of the world.

In her first letters to Derwent, Emeline confesses that she shares some of the romantic struggles and yearnings that Derwent is experiencing with Harold:

> You see I understand so well what you lads are feeling because I've experienced it myself. Every one of my intimate friends I have had to let go out of my daily life and I had to give up someone who was dearer to me—by far—than my own life. So I know all the agonies of parting.

We note that Emeline does not yet, in 1916, gender this person who she values more than her own life but had to give up for undisclosed reasons.

At this point, early in their decades-long relationship, Harold has gone on one of his first trips away from Derwent, with something in their parting leaving Derwent unsure and torn up about the status of their relationship. Emeline reassures him, a role which she plays frequently:

> But you have too much of the feeling that you are giving each other up forever and you are not doing that—you couldn't—you are a part of each other. Have faith in each other and in yourselves and the strength of an affection like yours and believe that you are being separated only for a season and trust life is bound to bring you together again. Each of you has helped the other: in the time you have been together, each has grown. Now the law of growth plainly requires that you separate for a time. Yield to the Higher Power that is shaping your life and trust to it.[10]

According to Emeline, Derwent has uprooted Harold from his previous life, a move that Emeline and Harold's friends support. Derwent is experiencing some kind of struggle and has gone away—neither to Chicago nor to New York, perhaps away doing research—with plans to return to Chicago to take up his work at the University of Chicago in September of 1916. Emeline not only reassures Derwent about the status of his relationship but encourages him to remember that change is an important part of the course of life: "Rise with it then! Rejoice that man is hurled from change to change unceasingly, His soul's wings never furled!" Emeline is quoting poet Robert Browning's "Under the Cliff," which she does at greater length in this same letter.[11] Emeline continues: "If it so happens that Harold does not come back from New York," and she then encourages Derwent, as usual, to find a way to come to visit her in Wooster.[12]

Emeline reveals herself to be invested in the romantic relationship between the two men, which she writes about as "a close friendship" between Harold and Derwent. She wants to know what happens, perhaps fearful of losing her friendship with Derwent, if Harold's relationship with him ends. She entreats Derwent to visit on his own: "Seems to me I can hardly stand it if I don't see one of you after things are settled. You see I have a great big heart-ache over it all, myself."[13]

Only two months later, when Derwent finds himself presented with an important work opportunity, likely to take up a faculty position at the University of Chicago, Emeline assumes a role that she will repeat throughout their relationship, as advisor and mentor. She urges Derwent to pursue this opportunity, not only for professional reasons but for the responsibility of supporting Harold:

> Of course you and I and Harold all know that no sacrifice you could make for him would be too great, provided it were for the good of you both. When you entered into this close friendship with Harold, you did assume responsibilities which leave you no longer free, as you were before. Friendship always has to be paid for with self-sacrifice and pain, like everything else in life that is worth while, but there is a certain point in self sacrifice beyond which no one has a right to go.[14]

At this point of uncertainty in their relationship and with Harold away, Emeline encourages Derwent to visit her on his way home from Newark, New Jersey (likely his arrival point from travel abroad), to Chicago, which he does. Subsequent correspondence in September reveals that this visit was an important bonding moment between Derwent and Emeline, particularly a conversation they had on Sunday night, which was altering for her, and none of which would have been possible had Harold returned when he was supposed to from New York and been present as well during their visit.

Emeline's unabashed, giddy enthusiasm for Derwent's visits and her insistence that he stop in Wooster and stay longer never abates. It is clear from their correspondence that they can affirm each other's romantic love and intellectual and cultural pursuits as a deeply intimate friendship unfolds.

Over time, Emeline outs herself as queer—albeit subtly, without identifying language or direct statements—with reference to her shared life and home with a romantic partner. Over the years of correspondence, she begins to write of Edith Yoder, who lives with her, wakes up with her, and—when Emeline's eyesight is failing in the mid-1950s—writes the letters that Emeline dictates, occasionally inserting a humorous little comment or facial expression, and always ending with the formal "per/EY" to indicate her role in transcription.

TRAVEL AND VISITS

Emeline and Derwent correspond frequently about their travel. Most often, they discuss Derwent's professional and personal travel to distant locations, from African to South American countries. Emeline travels on occasion, too, and their correspondence touches on these journeys. As Emeline notes in 1945, likely alluding to the physical challenges that compromise her mobility and ability to travel: "Considering everything, I really have got around in my life, haven't I?"[15]

Emeline also encourages Harold to spend time with her. In 1935, she spends Christmas with Harold at The Loft in Cambridge, and she thanks Derwent for preparing the apartment for their visit before he left on a trip to Latin America to travel and study Spanish. During this holiday in Cambridge, Emeline and Harold plan a trip to France together in 1936. In January, she writes to Derwent about these plans: "Isn't the plan for the trip to France exciting? I'd love to stop teaching now and read and learn for the next five months. I know so woefully little about the geography, history, and architecture. But, being a working-woman, I must earn my trip. You would be appalled by the heaps of blue books covering my desk."[16]

ON DERWENT'S WORK, WORKLOAD, AND WORK ETHIC

Emeline comments frequently reflect a knowledgeable observer—herself a faculty member—on Derwent's considerable workload and work ethic. Once the war started, for example, she wrote: "Derwent, the amount of work you get through is simply staggering; all those important lectures and this writing of books and articles. Did you finish the book on geopolitics? And now you have the Environmental Backgrounds and Agricultural Regions of the world on your hands! Or are you by chance in Washington again? They must be needing much work done there by men who know world geography."[17]

At one point, Emeline compared Derwent's workload to that of another hard-working colleague in her midst, stating, "How can you be any busier than last year—or the year before! You and Lewis are just alike. Every time he comes he said 'I have never been so busy in my life,' and yet you both seem to be always working to your full capacity."[18]

Sometimes Emeline requests work-related information from Derwent for her own teaching: "Der, please give me a definition of geopolitics which I can give to my students. I have an idea of course what it means, but can't get it into a few words."[19]

Emeline often congratulates Derwent on his achievements, noting his own modesty each time, such as when he is awarded a medal for his work in 1948, but does not explain it to her.[20] Similarly, in 1939, Derwent is building the Geography Program at Harvard, writing a book, and organizing the annual meeting of the Association of American Geographers (AAG), the discipline's main professional association in the United States. Emeline writes to him about all of this work (noting that Harold is away in Europe for an extended visit): "How fine it is that the geography program is building up so well; but it ought to, with you and Harold at the helm. Who takes his place this year?"

She continues:

I can imagine how much time and energy it took to plan and put through the meeting of geographers. After she returned, Helen Strong was good enough to write me an account, with an enthusiastic description of the high spots, not forgetting the A A G dinner and the Chicago dinner at your apartment. She said everything was perfect. Wasn't it thoughtful of Helen to do that? I was so pleased to hear the account. Yours would be too modest, I know.

Not missing any part of Derwent's work program, Emeline then turns from the conference to the book:

How about the book Derwent? You said it had to be at the publishers by Dec. 1. Did you succeed in meeting the date, and are you busy reading proofs, or don't you do that? When will it be out and who are the publishers?[21]

In this letter, Emeline also shares news of her own ill health and the difficult year she is having teaching due to this illness. She marvels at Derwent's ability and willingness to care for "Mrs. Kemp," Harold's mother, rather than return her to her "hardhearted" caregivers in Newark, while Harold is abroad for an extended period in Europe. We wonder if Derwent might have had a soft spot for Mrs. Kemp, not only because she was his partner's mother but she recognized and supported her son's relationship with

Derwent. This letter, like others, provides glimpses into Derwent's work ethic in his professional and personal life, and to his character and devotion to loved ones, with modesty and integrity. He cared deeply and extensively for those around him, whether hosting "perfect" professional dinners at the annual conference in Chicago or caring for his partner's mother while his partner is abroad, all while building a new program and completing a book.

Emeline affirms the importance of Derwent's work and shares her pride of him, boosting his spirits, especially when his self-esteem was low. She writes to him in 1943 after reading his book (which he must have turned in to the publisher eventually) and a published review of the book:

> The amount of important work you have to do appalls me. But aren't you happy to be working in a field so necessary to the success of the war. You know, Der, I am terribly proud of you, of being an intimate friend of "one of the most distinguished geographers of the United States." I got the Sat. Rev. of Feb. and enjoyed the appreciative review.

Emeline continues: "The clear conciseness of your book is just what I needed: concise yet complete. I feel that I could explain with a reasonable amount of clearness what geopolitics is."[22] This evolution of Emeline's understanding of geopolitics, from her earlier request that Derwent define the field to assist with her teaching to her ability to teach it after reading his book, is utterly endearing, especially to two political geographers tasked routinely with the challenge of teaching geopolitics. Emeline speaks in myriad ways, both personal and professional, of the success of Whittlesey's scholarship.

In a series of letters written in 1943, Emeline discusses how Derwent's "war work" is winding down, and she describes the effects of military personnel returning to Wooster, literally crowding the women students off the sidewalks and into tall snowbanks as naval cadettes walk in formation between classes.

During this time, she writes repeatedly about Derwent's book and the published reviews she is also reading (for which she advises him to hire "a clipping bureau"). She closes: "Ever so much love, Derwent, and the best of successes in your many tasks, a success which can't help but come because it is based on not only brilliant capacities but the power of unremitting, steady hard work, other than which there is hardly any quality I admire more. Yours, Emeline."[23]

In 1944, Emeline responds to letters that Derwent has sent describing his work trips to Montreal to give a lecture and to New York to serve on a panel. Emeline tries in various ways, yet in vain, to find the radio station out of Montreal to hear the public lecture, which must have been broadcast. Reflecting on the panel in New York, she writes, "It gives me the greatest thrill, Derwent, to watch your constant climb in the geographic world. . . . It must be immensely satisfying to feel you have so much to contribute, so much which few men have."[24] She continues, noting the reprinting of his book: "A new printing of *Earth and State*!"

Some years later, in 1952, Emeline writes about a book that Derwent decided to write with a former student, noting the importance of demanding work to herself and to her correspondent:

> I was happy to hear of the book planned in collaboration with your former pupil. It is a goal to strain toward, something attainable which demands your best. That is what we all need for a really rich and happy life.[25]

In reading Emeline's letters, we were struck often by the distinction between her prideful praise and the ways that scholars have written about and memorialized Whittlesey (e.g., Smith 1987; Martin 1988) as not the right person to defend and preserve geography at Harvard. While Emeline's views on her close friend are subjective, to be sure, the events she writes about are objectively impressive forms of public recognition and celebration demonstrating the importance of Whittlesey's scholarship and his professional role in the discipline of geography and in training a new generation of geography students. Whittlesey travels often to speak and publishes work that finds a public audience and high praise in the *Saturday Review of Books*. The book, published and reviewed during World War II becomes central to military and geographic education in the United States.[26]

EMELINE ON EDWARD ACKERMAN, QUEER HOUSEHOLDS, AND QUEER RELATIONSHIPS

While Emeline affirms Derwent's hard work and achievements, she also reminds him of the importance of living a full and balanced life, encouraging him to spend time with family: "I am very happy that you had the happy visit at home with your own people. They grow dearer as life goes on, don't they?"[27] Emeline also encourages Derwent to take time off so that he does not collapse under the weight of so much work: "Derwent, do cut out enough things so that you won't have to hold yourself together 'by nervous force.' So many men I know are crashing under the stress of the times."[28] Harold remains away at this time for an extended period in Europe, and Emeline encourages Derwent to enjoy his time at home alone: "Take advantage of Harold's absence to have some restful time at home. The apartment can't be so colorful with Harold absent but I am sure it is much more tranquil."[29]

Some of the drama and tense episodes between Harold and Derwent indeed involve the "colorful" nature of Harold and, specifically, Harold's relationships with other men. Emeline refers to these directly, sometimes with a tone of "Harold will be Harold," sowing his wild oats and having his drama. When she visits Harold in France, for example, in 1936, Derwent is concerned for her, and she writes from France explaining that he need not worry: she understands that Harold will do his thing, and she will entertain herself.[30]

On April 22, 1954, another drama came to an end, likely in which Harold was having an extended relationship with a man named Michael. It was not uncommon for men attracted to other men to ultimately comply with and perform society's heteronormative scripts and expectations, made all the more punitive during the Cold War McCarthy era in which the Secret Court rooted out homosexuality on Harvard's campus and the US federal government expunged queer people from its bureaucratic ranks (Johnson 2006; Wright 2006; Graves 2009).

Michael eventually reports to Harold that he will marry a woman. Emeline characterizes Harold as "so miserable," expressing her sympathy for Derwent, and writes: "The news about Michael was not surprising, because he was bound to pull away from Harold sooner or later. I am glad for Michael that he is to have a wife and a home, a normal life. He probably *didn't* tell Harold about it before H. went over and during his visit because he was afraid to. I think he should have told Harold when he first got interested in a girl."[31]

Emeline goes on to recount Harold's ensuing depression, which is so difficult that it affects Derwent's ability to write and required Harold to see a psychiatrist. Emeline attributes Harold's "neurotics" to the "selfish and disorganized life of his family," while also noting his own self-absorption: "It's such a pity that Harold has always been so possessive in his loves and can't rejoice in other people's happiness."[32]

Among the men narrated frequently as important in the lives of Harold Kemp and Derwent Whittlesey is Loft-mate Edward Ackerman. While occasional inferences arise to Harold's fondness for Edward, above all, Emeline recognizes him as an essential member of their queer household. When Ed leaves in 1942 to work in service of the war, Emeline writes:

I sadly fear that Ed is gone for the war period. Of course I realize to a certain extent how great a vacancy his going leaves in the household. But aren't we thankful he can remain out of the army! . . . Ed meant so much to the happiness of your household. I feel that he'll come back to you, though. He is extremely loyal and you people are entirely his family. Don't worry about his not ultimately returning.[33]

Six years later, after Ackerman's return from the war, all three men once again must confront another departure from Cambridge. Emeline comments once again on Ed and his place in their home: "How fine it must be to have Ed back again. I can imagine the endless stories he has to tell. It Is distressing to hear that he is going to leave Harvard. I should think they would do anything to keep him. You'll have to find some other nice young man to come in with you two."[34]

History tells us that Edward, too, eventually marries a woman with whom he has four children. His departure from Harvard—and, therefore, Cambridge and The Loft shared by this triad—leaves Derwent, Harold, and Edward in deep sadness and grief, which we will discuss in subsequent chapters.

EMELINE'S FRIENDSHIP, AND THE PASSIONS AND JEALOUSIES BETWEEN DERWENT, HAROLD, AND EDWARD

For four decades, Emeline plays an important role in Derwent's life, and Derwent in her own. They affirm each other as friends, as partners, and in work with deep mutual respect and understanding. Although Harold is Emeline's cousin, she often articulates for Derwent how difficult it must be to weather Harold's mental health struggles, stormy emotions, and challenging behaviors, from that very first letter in 1916, when Derwent is in turmoil about Harold's solo trip to New York and whether he will return to Chicago. In 1943, three decades later, Emeline writes, "It was splendid news to hear that Harold is to have the same sort of course as at Columbia and that he has finally shaken off his sadness. These emotional entangle-ments are hard to handle."[35] At another point, a letter arrives that carries no specific date and is labeled only "Easter Sunday," but the contents relay it is written seven years after Derwent has moved to Cambridge, so likely around 1935.[36] At this time, as noted in chapter 2, Derwent is defending Harold at work by trying to advance his position beyond one-year con-tracts and threatening to resign if Harvard's administration does not agree.

Meanwhile, Derwent and Harold are having a difficult time in their relationship, and Derwent is living with duress that is hindering his work, friendships, and well-being. Emeline finds time to write Derwent a long letter as she nears the end of her week of vacation visiting a close friend's daughter in Ravenna, "in the direction of Youngstown," Ohio, as she explains at the letter's opening.

This long letter is impassioned in her response to emotionally distressing correspondence she has received from Derwent and to which she responds, first affirming her care and love:

> I wonder what you are doing today, and what Harold is doing. You two are much on my mind. It is literally true that I think of you every day and long to see your situation alleviated. Of course there is nothing in the world I can offer except understanding and deep concern. If there is any solution to the problem you must find it yourself. The only solution I see is one to which you have so far been unable to come, but which you will one day arrive at.[37]

> I have read your letter over several times. It is very evident that Harold is[sic] neurosis is much worse; it is also evident that your endurance is approaching its limits. In common parlance the affair is getting on your nerves. You have a very serious problem, not only one affecting your personal happiness, but the quality of your work. You are approaching it too much from the emotional angle. Try to do some clear, crisp, dispassionate thinking about it. Through a strong attraction, much congeniality of taste, and long and intimate dissociation you are closely bound to some-one with whom it is increasingly difficult for you to live and keep that spiritual free-dom and self respect which every balanced individual must have. Love suffereth long and is kind but love must not suffer indignities and humiliation. How can you let Harold bring you to the point of beginning to believe that you have nothing better than a pedant's brain? Try to look at the matter with the intellectual clarity you would bring to bear on a geographical problem.[38]

Emeline's response is an interesting mix of compassion and dispassionate pragmatism, masculinist in its dissociative approach to Derwent's emotions here, suggesting that he simply bring "intellectual clarity" as if to any "geo-graphical problem."

Emeline continues:

You have reached a position very enviable for a man still young and you would be further along now, had you met with someone in your personal life who would give you encouragement and inspiration instead of paralysing you with cruel criticism! Where would Harold be now, had it not been for you? He never would have finished even the work for his bachelor's and you know it. Probably he does too.[39]

Emeline pleas with Derwent to move out and live independently:

It would be a very sad thing if this tie of many years should be completely broken. Before that is done, I should certainly think you would try living far enough apart from Harold to preclude any daily intercourse. Cast your mind back to the state of your relationship when he was at Dartmouth or you were living in the student house. Was it better? If so, why don't you give up your present apartment in June and with fall take one farther away, no matter what he says about it?[40]

While encouraging distance, Emeline also addresses problems with conven-tional forms of monogamy:

If there is an enduring quality in your relationship that could not break it. That is the only suggestion I can make at present except one about your attitude toward other friends. Isn't it always a mistake, even in the marital relation, to stake all one's affection on one person? You have done that far more than was wise, but you still have plenty of time to retrieve this mistake. I am going to put all together the remarks in your letter on this subject, [now quoting Derwent's letter to her]: "a formidable problem—he has made it so clear that I have no friends but this [unclear] that I am convinced that no one likes me. Of course I have given up all the people except those whom he cares for—I have always been a one-man dog—Have no social initiative and always let a contact drop—One serious matter I have only five or perhaps six friends in all this city after nearly seven years of residence here. And not one of them is my own age. Will you tell me what is wrong? . . . I don't relish the prospect of lonely old age, and if I ever had the capacity for making friends, it has atrophied."[41]

Returning, now in her own voice, Emeline explains:

Maybe I seem a bit cruel, Der, but I am very deeply and vitally interested in this thing. I am your real friend. You can't get along without friends and you are going to find it increasingly lonely and unsatisfying without them. . . . You know perfectly well that no capacity has atrophied in a man in his forties, and you are still in your early forties, aren't you? Five or six friends is a grand start. So far as age goes, friendship is more or less an ageless thing. You have so much to offer in a friendship. I'll repeat what I said: a brilliant mind, an attractive personality, much knowledge of various forms of culture, much information about interesting places, delightful powers of conversation, delicacy and sensitivity in spiritual relations.[42]

She then urges Derwent to "cultivate what you have," to "deliberately go after any people to whom you are slightly attracted and see what happens. Make some experiments. For heaven's sake don't sit down and say, 'The Lord made me so and so there nothing to be done about it.'"[43] Emeline recognizes the abusive manner in which Harold is insulting Derwent and attempting to isolate him from friends, well-established tactics in emotionally abusive relationships. She continues, "What does Harold think is going to happen to your relationship if he continues his present course? He has been acting this way for so many years that I suppose he thinks he can until the end of the chapter. I remember so well hearing Lewis tell of the time he ran off and hid for many hours."[44]

In another letter, which we believe was written years later, Emeline addresses this same emotional turmoil and encourages Derwent to move in even stronger terms as the distressing situation continues:

> It was very good of you to write me the long letter telling me of the situation with yourself and Harold. It makes me just heart-sick to hear how bad things are and the hopeless thing is, that these extreme neurotics never improve, but grow steadily worse. There is only one thing for you to do—move far enough away to not be at Harold's beck and call and live your own life. That seems impossible, I know, but I am sure that if the break were once made you would experience an enormous feeling of relief and that the joy of life would gradually return. You have everything to make life delightful—a brilliant mind, a charming personality, a fine position and real love for your work. How shocking that the selfish and jealous caprices of another human being render all this dust and ashes. I can't think about it too much for my blood begins to boil.[45]

It is important to note Emeline's support of her dear friend, who is not a "blood" relation, but a queer kind of family, over and above his partner, her familial cousin, whom she speaks of in terms that address mental health struggles and possible emotional abuse:

Of course, I fully realize how your anxiety as to Harold's health complicates the situation. The threats of suicide mean nothing. I have heard his father utter them dozens of times; that was the club he used to hold over our cousin Jenine and he died in his bed at the age of eighty-seven, wasn't it? All these emotional orgies would do something to a heart, one would think, yet Harold may live to put flowers on your grave and mine![46]

After her letter takes a morbid, if premonitory turn, for Harold—though senior in age—will ultimately outlive Derwent and Emeline, she turns to queer humor and encourages him again to move out to get away from Harold:

You say that all the women who are interested in you think that you ought not to live together but do not say why. My reason is of the simplest? You ought not to be a slave and he ought not to have the opportunity to vent his selfish emotions on you. If you are going to live in the same street, you had better live together for financial reasons, if for no other. To be in a different apartment but within call is no alleviation at all.[47]

She continues in even stronger terms:

How I wish I could suggest a *modus vivendi* but I am convinced there is none. How can you ever live happily with such a person? I am shocked to hear how really awful the situation is. To be constantly in this state of mental and emotional turmoil is ruinous to work.[48]

She affords Harold some support:

Do not think that I am un-sympathetic to Harold. This letter does sound cold and hard, I know. I know all his fine qualities and admire them and I feel deeply grieved over his misery. Can you imagine how fearfully unhappy anyone must be who is in his state of mind? Of course he knows that you are the best, the most patient and most sacrificing friend he has ever had. I sometimes wonder what would happen if he realized that he had to behave himself or lose you altogether. Has not he always felt too sure of the tie that binds you to him?[49]

Emeline then returns to the issue of jealousy in a way that affirms Harold's affection for other men, and specifically, his affection for Edward Ackerman, as well as her own relationships with women:

What a terrible monster jealousy is! I hope Ed will never marry and live anywhere near Harold. I have had a good taste of the effects of this passion this year. Adele has been outrageously jealous of the girl who drove me last summer. Adele herself as been too nervous to drive yet she staged a scene every time Florence drove for me. The result was that I used my car very, very little this winter.[50]

Emeline closes this revealing letter with resignation:

This is a very unsatisfactory letter, Derwent, but I do feel so helpless. If we could only talk we might thresh things out a little more completely. If you could only harden your heart! It might be better even for Harold in the end if you would leave him to work out his own life; but it is hard to know.[51]

Ultimately, though, Harold and Derwent stay together, sometimes finding outlets with other men, as in Harold's relationship with Michael, which ends in an emotional crisis for Harold and, by association, Derwent.[52] Emeline writes, "I wish Harold would realize what a precious gift he has in your loyal friendship."[53]

DERWENT'S FRIENDSHIP SUSTAINS EMELINE AS SHE STRUGGLES WITH HER HEALTH

Derwent, reciprocally, respects and understands Emeline and demonstrates deep empathy and awareness of her health struggles. Emeline writes often about ailments that cause her to miss classes she is scheduled to teach, instead sending her to bed for weeks at a time.

Derwent proves himself a generous and doting friend. In addition to his steady correspondence, Derwent frequently sends Emeline gifts, like books, music, and chocolates, for which she writes effusive thank-you letters. Some gifts are intended to alleviate some aspect of Emeline's suffering from poor health. In January of 1943, she thanks Derwent for one such gift, hand-crafted and intended, thoughtfully, to alleviate the arthritic pain that prevents her from achieving mundane tasks:

How very, very good you are to me! The tongs, my picker-uppers, I call them, are *wonderful*—perfect for my need and solve a most annoying problem. You would have laughed to see me throwing hair pins, bobbies and clips on the floor and picking them up with child-like glee. Thank you, Derwent! But *where* did you find them? They are evidently hand-made. And who wrapped them up in those ancient newspapers of fifty years ago?[54]

At this time, Emeline is largely confined to bed and a wheelchair, though speaks of some improvement that enables her to walk once a day to the living room to sit on her couch for an hour before she suffers another fall and injury. At this point she observes, "If there ever was a collector of accidents, I am *it*!"[55] Derwent offers to send her records, which Edward has, because—in addition to the small radio near her bed—she is enjoying a new record player that the Carnegie Foundation sent to universities, including the College of Wooster. However, she has limited access to records. Emeline also thanks Derwent for recounting his Saturday night visit to hear the Boston Symphony perform. After praising Derwent for his thoughtfulness and generosity, she closes this letter: "Much love, Derwent. You are such a good friend. Emeline."[56]

Derwent's friendship and correspondence sustain Emeline, just as hers sustain him, and she never loses her sense of optimism. After detailing a difficult series of ailments, she notes that a "a most evil bug found me and sent me right back to bed . . . the trough of the wave," she writes. "The zero weather can't last much longer, can it, because the earth does turn and spring has to come."[57]

In January of 1944, Emeline writes of the growing importance of friends, imploring Derwent to come for a visit. She laments how poor Harold is at keeping in touch and visiting and writes, "My greatest pleasure now is seeing my friends. I have no life of my own now. I live completely in the joys and sorrow of those I love, of whom you are one of the very dear. You are such a *good* friend to me, Derwent."[58] Although mostly confined to bed after a recent hospital visit for undisclosed reasons, Emeline wears a brace to move around her house a little and sit in a wheelchair for one hour each day. She entertains a steady stream of visitors, "an average of two or three callers a day," who keep her abreast of college affairs.[59] She also recounts books she has read and debated and others she plans to read, records she listens to with her companion for an hour in the morning and an hour before bed, and politics.

Emeline has a coterie of young men whom she looks after, hosts, and whose lives and careers she follows and supports. Most seem to be part of her extended family. One who appears frequently with reports in correspondence is Paul, who becomes an academic, and who—like most young men—enlists in military service during the war.

Emeline and Derwent each share in supporting one another through life and work. Emeline sustains Derwent through relationship struggles and the most devastating crisis of his professional career, when his program comes under threat of closure. She worries frequently over Derwent's unhappiness after this time, responding here to one of his letters during this difficult period: "Of course work is the great solace. I am glad that you have so much of it to do. However you should have a vacation somewhere. I am still hoping you will see fit to come out to Ohio, just to see me!"[60] She ends this letter, "Do let me know how things go this summer. I *wish* you could be happy. Much love, Emeline."[61]

Emeline is ailing and retires from teaching by the mid-1950s. She has trouble getting out and relies heavily on Edith. Edith transcribes Emeline's letters, adding lines to the end in her own voice, to report on Emeline. In 1954, she describes bringing Emeline onto the porch for two hours to watch her plant vegetables in the "postage stamp of a garden." She also comments on the pleasures Emeline continues to derive from music, being read to aloud, and from Derwent's correspondence: "Miss McSweeney enjoys your letters so much, I hope you will always take a little time out to write. It means so much to her."[62]

"DO WRITE ME AGAIN SOON, DERWENT": EMELINE'S FINAL LETTER

What turns out to likely be Emeline's final letter to Derwent arrives in September of 1956, about two months before he dies. It is a satisfyingly long and newsy letter, dictated to and transcribed by Edith, the woman we infer to be Emeline's partner and living with her at the time. In his meticulous dedication to the preservation of their correspondence, Derwent files this letter alongside all the others, spanning the years, life events, and emotional intimacies they share and that stretch to connect them across the distance since that first letter arrived in Cambridge in 1916. Derwent answers this letter, a fact he reports in his own handwritten note in the upper-left-hand corner of the front page: "Answered Sept 16, 1956." Emeline opens, "Dear Derwent, I was delighted to receive your letter of Aug. 5th. Never say that your letter 'must seem dull.' Because no letters from dear friends are dull."[63]

In this last letter, written—albeit unwittingly—so close to the end of Derwent's life, Emeline once again engages with the routine news of Harold and Derwent's shared day-to-day lives as she finds herself ailing alongside Edith in her own life: "You say that you and Harold have had a very quiet summer, but teaching ten hours a week and attending all that theatre sounds like a very exciting life to me in my bed. Things are relative you see."[64]

She is pleased to hear that Derwent and Harold have just purchased a "magnificent new car!" Speaking of the comforts of home and a life confined to bed, Emeline describes enjoying the new "grand air conditioner" given to her by an affluent former student. She recounts that lightning struck during a recent electric storm, destroying her old telephone that had been in the dining room, and prompting a move of a new telephone into her bedroom where she can now more readily hear it ring and answer.

Emeline reports on aspects of aging, from time in bed to not hearing the phone and compromised vision: "My eyes are failing gradually so I have to use a magnifying glass for most reading—consequence, much less reading than I like to do. I listened a good deal to the Conventions and don't be too shocked—I've fallen back on a few soap operas!"[65] We learn, eventually, at the end of the letter, that Edith is writing on behalf of Emeline. She reports on her daily life with Edith: "Of course we are mightily interested in the big Kroger Super Market going up near us"—a supermarket that still stands next to the house at 659 Beall Avenue.

Here, in the final lines of the last two pages of this letter, Emeline most definitively reveals herself to her unforeseen readers as queer, in the detailing of a shared intimate life with Edith: "Edith came back last night after two days. . . . Saturday she attended a big wedding where she saw many of her North Dakota friends here for the occasion. Yesterday the Yoder clan some 25 in all had a picnic at the city park with mountains of food, consequently she is a bit groggy this morning and I'm really punishing her by having her write this letter." This last line is followed by a frowning smiley face rendered by Edith herself. Emeline thanks Derwent for sending her an article on Rhodesia, which she was reading with her magnifying glass and will donate to the library upon completion. Emeline signs off as follows, with oft-repeated sentiments, and after Harold has suffered a heart attack and is recovering: "I hope Harold continues to gain strength. Much love to you both. Emeline," followed by "per E.Y."[66]

We can only imagine the pain that Emeline must have felt upon learning of Derwent's sudden death, after sending her final letter. We can only wonder if she would have been well enough to travel to attend his funeral, joining Edward and Harold in Cambridge to comfort them in the immediate shock of grieving and dealing with logistics, if her health and circumstances allowed.

We inquired a second time, in 2023, with librarians and archivists at the College of Wooster, in a hopeful yet ultimately vain attempt to find any records of Emeline's that might have been kept archived in Special Collections. Emeline died nearly two years after Derwent, on November 5, 1958. Our second inquiry turned up additional photographs (figure 3.2). After this final letter in Derwent's file, the only record we have of Emeline is a brief obituary published by her sorority, Kappa Kappa Gamma, two years later, recognizing that she became Professor *Emerita* of French after forty years of teaching, first in the preparatory department and then at the College (Kappa Kappa Gamma, 1959, 44). For this service, the sorority (at the time called a fraternity) had honored Emeline—"widely known teacher and educator, long a member of the Wooster College Faculty"—with an educational award, four years before her death (Kappa Kappa Gamma, 1954: 103). Figure 3.5 shows the article announcing this award, published with Emeline's photograph above the announcement.

Four receive new educational awards

Miss Sweeny   Miss McSweeney   Mrs. Rice   Mrs. Smart

Kappa announced the presentation for the first time of four special educational honor awards.

They went to *Emeline McSweeney*, B Γ-Wooster, widely known teacher and educator, long a member of the Wooster College faculty. Miss McSweeney is now retired but still makes her home in Wooster, Ohio.

Lucy Allen Smart, B N-Ohio State, a former editor of THE KEY and other organizational bulletins is dean and librarian of Kew Forest School in Forest Hills, Long Island. For a number of years Mrs. Smart wrote and gave

missionary service in Forman Christian College, Pakistan and Allahabad Christian College, Allahabad, India. Both Dr. Rice, who is president of Forman College, and Mrs. Rice are now on furlough, working in the office of the Presbyterian Board of Missionaries in Chicago. Her particular interest is with foreign students in and around Chicago.

The last of the honor awards went to *Mary E. Sweeny*, B X-Kentucky, former assistant director of the Merrill Palmer School, Detroit, Michigan and noted authority on home economics. Widely known for her study

Figure 3.5
Photograph of Emeline McSweeney (second from the left) in Kappa Kappa Gamma's *The Key* when she wins educational award, 1954. Source: *The Key*, Kappa Kappa Gamma.

CONCLUSIONS

Emeline's personal letters stand as their own remarkable queer archive, nestled within a larger masculine, white queer archive housed in a powerful, elite institutional setting where other such stories that would surely threaten one's employment status very likely disappeared, never to be shared. In this chapter, we discussed Emeline's correspondence and, in so doing, complicated heterosexual, masculine histories of the Geography Program by introducing its intimate queer and feminine contours and documenting what we know of Kemp and Whittlesey's personal responses to the threatened closure to the Geography Program, mediated through McSweeney's compassionate written voice.

Emeline frequently observes Derwent's professional successes—books, awards, and other achievements—and his exceptional work ethic. As their correspondence highlights different periods of struggle for Whittlesey and Kemp, a chain of letters from 1948 to 1949 reveals the most intimate feelings about the intertwined fates of their lives and the Geography Program at Harvard. Emeline supports Derwent when the potential closure of the Department of Geography threatens to undo three decades of his work.

Ultimately, Emeline's support of Derwent strengthened him to resist the administration's actions from then until his untimely death in 1956. Her letters document the widespread resistance to the closure of the Department of Geography and the role that Derwent played in that resistance. In so doing, they counter persistent narratives in the existing record that portray Whittlesey as too passive a person to fight for geography (e.g., Smith 1987; Martin 1988; Wright and Koch 2009). In a story otherwise dominated by white, heterosexual masculinity, Emeline McSweeney's haunting narratives in these files belies the notable exclusion of the queer and feminine in histories of Harvard's Geography Program. McSweeney, therefore, functions as an important figure in the queer archive: one where queer love can be spoken, where life's trials can be shared openly, where love, acceptance, and refuge from the homophobic storm of institution and society is offered.

Emeline often wrote at length about books and music she had recently discovered. As a story within our story, in February of 1945, she writes to Derwent to thank him for two novels he sent her by author Georges Simenon, including *Affairs of Destiny*. She explains that these occupied her nightly uninterrupted reading hours from 10 p.m. to midnight for several nights. Emeline's description of *Affairs of Destiny* lends the book she discusses in her letter—which we now discuss in our book—a premonitory quality:

> *Affairs of Destiny* I found very absorbing for several evenings. They are fine psychological studies. In the first, the way in which the hero enveloped himself in his own destiny reminded me of a spider weaving itself into an inextricable web spun from its own body. *The Woman of the Grey House* has a stark, stripped sort of horror about it that is ghastly. Thank you for thinking of me, as you so often and generously do, my dear Derwent.[67]

A review of these two short novels, published in the *New York Times* the year prior, also sets the two apart: one about a man consumed by his indecision until he becomes the murderer he hesitated and failed to report; the other about a woman who kills with no compunction, her head held high (Anderson 1944). The uncanny nature of these stories is that they parallel both our story and *its* haunting: the innocent transformed into the guilty, the responsible acting without the burden of guilt, and the subjective, spectral, and agentic spun into the objective, material, and fated, just as a spider becomes a living part of an otherwise inert web.

Reading letters that wound their way slowly from Emeline's home in Wooster, Ohio, to The Loft on Prescott Street in Cambridge reflected and humanized the institutional events we detail in the chapters ahead, showing how they affected both Derwent and Harold. Emeline speaks back to Derwent and Harold about their experiences. She mirrors, but refracts as well as reflects, their context and their lived experiences and emotional responses, at times encouraging and at others challenging their status quo, giving an important and distinct voice to this narrative. Together, Emeline and Derwent's letters spin a web that connects them emotionally in their daily lives, through their beautiful, detailed correspondence.

A queer woman's voice shared in letters that spanned four decades—most of Derwent's adult life, from his twenties in Chicago to his sudden death days after turning sixty-seven in 1956—also reveals much of her own life as she works, writes, ages, and grows in Wooster, as a part of the close friendship and intimate correspondence between two people who met, understood, and witnessed each other's growth and lives over many years.

In their correspondence, intimate relations and relationships were given space to breathe and exist. The complexity of queer relationships was explored in these handwritten and typed pages, the powerful emotions of an illicit and hidden love affirmed and validated. Queer relationships include earlier language used to identify same-sex attraction, or attractions other than heterosexual between masculine and feminine identifying, and gender-queer people. Their correspondence references Uranianism, for example, a term that referred to queer people in the late 1800s and shortly into the turn of the century, as well as an associated intellectual movement at the turn of the century (Pettis 2004). In addition to a shared language, they share cultural references in arts and literature.

In his penultimate year of life, Derwent typed a letter to a colleague with a son living in Wooster. He wrote the following: "Did your son ever meet in Wooster the remarkable Emeline McSweeney? I never saw a more extraordinary character. She has been lame all her life, and has had a succession of devastating physical ills—now at nearly 80 she still has one of the liveliest minds I have ever run across."[68] The relationship between Emeline and Derwent was clearly fueled as much by deep respect, admiration, and comradery as it was by love. Above all, the correspondence between Emeline and Derwent provides a beautifully written testament to queer love and queer friendship, a friendship that was loving and sustained across a lifetime.

THE RISE, FALL, AND "UNFINISHED BUSINESS"
OF GEOGRAPHY AT HARVARD

Although I had hoped for an explanatory statement from the College concerning the demise of the Geography Department, I now write not only in contention with the move but also to elicit some comment from those who made it.

Geography is important. Our recent World War demonstrated that to this generation. Geography as a science of the world's spatial relationships and patterns of human development is essential to the "Education for Citizenship" that many modern educators expound as the true aim of the Liberal Arts College.

I can only feel that the removal of the department is contrary to Harvard's General Education program and the newly organized, though less publicized, Regional Studies Plan, and to the best interests of Harvard in general. Before our valuable staff in the department takes leave of us, the edict should be carefully reconsidered. (Harvard Crimson March 9, 1948)

There are three theories [to the end of the Geography Program at Harvard]. . . . The third and most widely accepted theory is something uglier. In the mid-20th century, a number of Harvard's geography professors—including Whittlesey—were gay. The story goes that, to avoid openly firing the professors, Harvard quietly disbanded their department. The closure of Geography as a study at Harvard touched off a wave of similar moves at universities across the country; effectively, for some, it was the death of a field.

To this day, the circumstances around the end of the Geography Department remain shrouded in uncertainty. The excuse given [by the administration] to *The Crimson* in 1948 seems almost comically unspecific and dismissive. (Harvard Crimson April 21, 2016)

On January 13, 1948, the administration of President James B. Conant wrote to Provost Paul Buck of the Faculty of Arts and Sciences (FAS), raising the question of "whether Harvard University should proceed with geography or let the Department die on the retirement of Professor Whittlesey."[1] Later, after informal talks with a number of professors, Buck wrote his own letter to the chair of the Department of Geology and Geography, enclosing a copy of Conant's letter, where he stated, "It is now time to make a policy decision by saying that geography is one of the things that we will not develop at Harvard. . . . Harvard cannot hope to have strong departments in everything."[2] He also directed the chair to inform Edward Ackerman of this decision. Three days later, the department met and voted that no more undergraduates could be accepted as concentrators in geography, nor could the department accept new graduate students. As Harvard's newspaper, *The Crimson*, would put it nearly seventy years later, "geography was gone for good. Or so it seemed."

We opened this chapter by juxtaposing two quotations from student journalists writing for *The Crimson*. The first was penned by geography student Peter Roll in March 1948, the second by correspondent Emma Talkoff in 2016. While Roll's editorial directly assigns responsibility for the unfolding events to the Conant administration's "edict," Talkoff's article alludes to uncertainty regarding what happened, providing three theories—none of which she supports with any evidence or source. In effect, the beginning of the end of geography at Harvard, according to Talkoff, remains shrouded in mystery, even in contemporary times. In writing this piece without evidence, Talkoff takes license in a way that perpetuates myths with a false, if dramatic ending: the death of a field. The closure of the department at Harvard was equal parts distressing in its time and haunting in the present. The only location, moreover, where one could declare the death of a field that was, in fact, empirically growing and thriving across the United States at the time of publication would be by confining the research for the article to a campus which no longer had geographers.

But exactly *what* happened to geography at the college? What was the program's story? How could it continue to haunt Harvard, and even geography as a field, into the present?

Here, we delve into the longer institutional history of geography as a field of study at Harvard. In exploring geography's rise and fall at the college, we provide an empirically informed, more elaborate account of its story. We contextualize the situation by revealing more about this book's key characters, most notably, Derwent Whittlesey, who was at the center of building the program, suffered significantly once it ended, and fought hard for its preservation and hopeful restoration. The availability of archival sources that were not accessible to Rita Morris, Neil Smith, or Geoffrey Martin has led us to a more complex and different story than imagined by previous scholars. We recount this history in ways that prompt examination of the interplay of relevant social forces in the Geography Program's fate, such as homophobia, McCarthyism, and the politics of scientific legitimacy.

EARLY HISTORY OF GEOGRAPHY AND THE PHYSICAL GEOGRAPHY PROGRAM AT HARVARD UNIVERSITY, 1642 TO 1928

Geography began as a field of study in Harvard University's founding years, as early as 1642, in the field of "new astronomy" (Davis 1924; Morris 1962, 17). Rector Mather made geography a field of instruction in 1689, where it continued as a course of study until at least 1787 (Morris 1962, 52). During this time, geography became a core part of Harvard's second professorship in 1726: the Hollis Professor of Mathematics (Morris 1962, 18). Limited instruction in geography continued in the Department of Natural Philosophy until 1838, reappearing as a formal elective in the 1840s. Despite these presences, Morris (1962) showed how the field remained relatively limited at Harvard at this time.

The college reestablished and created more permanent instruction in geography beginning in the 1860s. We found a more regular record of the subject at Harvard through the president's annual reports to the Board of Overseers from 1826 to the present. These reports show a reemergence of geography as an undergraduate elective from 1841 to 1846 (Harvard University 1826–1995). Regular lectures, however, would begin in 1863 and continue until the program ended in 1958 (Harvard University 1826–1995).

During this time, American students were returning from Europe having learned the new discipline of physical geography, as taught by scholars like Carl Ritter, Alexander von Humboldt, and Friedrich Ratzel (Davis 1924; Morris 1962, 22). Instruction in physical geography also coincided with an "urgent demand" for professional geographers at the United States Geographical Society (USGS). Harvard concurrently created "Natural History" as a unified field of study, which encompassed, among other subjects, geology and geography (Harvard University 1826–1995). The university also required geographical knowledge for admission for the first time in 1869 (Harvard University 1821995).

After an upsurge in students and research, Harvard reorganized the Division of Natural History to include physical geography as a regular program of study (Morris 1962, 85). Infrastructure and material support for geography came from a specially allocated section of the newly built Agassiz Museum

Table 4.1

Timeline of key events in the history of geography at Harvard University, 1863 to 1962

Date	Event
1863	Geography becomes a regular undergraduate elective
1870	J. D. Whitney becomes the first professor of geography at Harvard
1888	Harvard completes a section of the "Geological Museum" to house geography
1896	Founding of the Department of Geology and Geography
1903	Peak of the Physical Geography Program
1904	Isaiah Bowman first appears at Harvard as a research assistant
1912	Local minimum in geography enrollments and courses as program weakens; W. M. Davis resigns
1916–1918	Revival of geography as Harvard becomes a center for training military officers
1926	Harvard offers Bowman a separate department and chair if he becomes professor of geography; he declines
1927	Raoul Blanchard is appointed professor of geography to head a Human Geography Program
1928	The FAS recruits Derwent Whittlesey as a second professor to build human geography
1932	The Institute for Geographical Exploration is founded with funds donated by Alexander Rice
1933	James B. Conant becomes Harvard's twenty-third president; Blanchard's last semester teaching on campus
1934	Harold Kemp becomes an instructor in the Human Geography Program
1936	Blanchard officially resigns, leaving Whittlesey in charge of the program
1937	Bowman stores a secret letter in his archives on his intense opposition to the Institute at Harvard
1938	Edward Ackerman is appointed assistant professor in the program
1941–1945	Geography at Harvard rapidly grows with wartime needs for officer training
1944	Whittlesey becomes president of the AAG
1944	The "Geography at Harvard" report recommends creation of a Department of Geography
1946–1947	Zenith of geography at Harvard with influx of GI Bill students
1947–1948	Three teaching staff in geography, including Ackerman, rejected for promotion, leading to their termination; Conant initiates "Let geography die" policy, leading to its removal as undergraduate field of concentration

Table 4.1
(continiued)

Date	Event
1948	Bowman appointed to Board of Overseers and tasked to geography; after review in October, the board calls for geography's immediate revival
1950	The FAS creates subcommittee on the issue, concludes department should be created
1951	The Institute for Geographical Exploration closes due to lack of funding; building turned over to Department of Mathematics
1952	LSE Professor Dudley Stamp conducts external review of geography, concludes Conant's decision should be overturned and geography revitalized using Institute's infrastructure
1954	Harvard committee on behavioral sciences declares geography's position as one of the university's "serious unresolved problems"; proposes to rebuild
1956	Whittlesey, last remaining geography professor, dies unexpectedly a few months prior to his planned retirement
1958–1959	Final geography courses that were originally part of the Geography Program taught at Harvard
1962	Department of Geology and Geography renamed "Department of Geological Sciences"

of Comparative Zoology in 1887, where the Geography Program would be located until it ended in 1956 (Harvard University 1826–1995). It was under these conditions that geography became a regular field of instruction in the natural sciences and housed in the museum in the late 1800s.

Geography achieved a number of firsts at Harvard by the end of the nineteenth century. J. D. Whitney became the first professor of geography in 1870 (Harvard University 1826–1995). William Davis taught the first physical geography course in 1878 (Harvard University 1826–1995). Geography became part of the new Department of Geology and Geography in 1896, conferred its first graduate degrees a year later, and began issuing annual reports in 1900 (Harvard University 1826–1995). Building on a previous map collection, the university opened a geography library in conjunction with Boston Public Library, New York Library, Columbia, and the John Crear Library of Chicago (Harvard University 1826–1995). Through its rapid growth and the presence of notable scholars, the department became a leading center of physical geography in North America by the early 1900s (Davis 1924; Morris 1962, 106–107).

Harvard's research and instruction in physical geography peaked in the
first decade of the 1900s and declined until the rise of human geography
in the late 1920s. The Physical Geography Program's enrollments peaked
in 1903, with 321 students enrolled in nine courses taught by two professors and four teaching fellows (Harvard University 1826–1995). As shown
in figure 4.1, this peak was followed by a sudden drop, with only thirty-
five students enrolled in three courses by 1912 (Harvard University 1826–
1995). Rita Morris (1962, 112–114) identified 1905 to 1907 as a period of
curtailment of geography at Harvard, even though the field was "emerging
by itself" in this period. Isaiah Bowman worked as a research assistant in
geography for W. M. Davis at this time (Harvard University 1826–1995).
Morris (1962, 118–119) argued that while the program succeeded, thanks
to Davis's presence, his resignation in 1912—along with the field's general decay in the 1910s—led to geography's temporary decline at Harvard.
The FAS, in tandem with the university's administration, reallocated new
resources and available professorships to geology instead (Morris 1962,
114). World War I temporarily abated this lull when Harvard became a

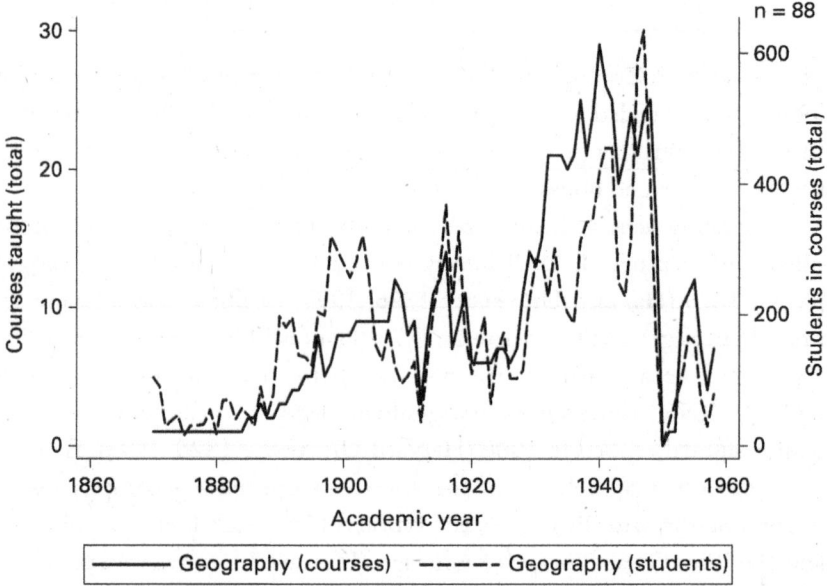

Figure 4.1
Frequency of student enrollments and courses in geography, 1870 to 1958. Source:
Harvard University Archives, Harvard/Radcliffe Annual Reports (1871–1959).

center for training military officers in geography, among other disciplines (Davis 1924), a pattern that was repeated with the surge in instruction and student interest in human geography during World War II (see Barnes 2006; Barnes and Farish 2006; Barnes 2016).

THE RISE OF THE HUMAN GEOGRAPHY PROGRAM AND DERWENT WHITTLESEY'S RECRUITMENT, 1928 TO 1948

Human geography began at Harvard in the late 1920s, emerging from instruction in regional and topical geography in and around the university in the early 1900s, with the creation of a formal program and the hiring of human geographers Raoul Blanchard and Derwent Whittlesey in 1928. Despite the Department of Geology and Geography's focus on physical geography in the first two decades of the twentieth century, what we would now identify as human geography was taught elsewhere on campus. Early examples include instruction of economic geography in the Harvard Business School and "Political Geography of Europe" in the Department of History (Harvard University 1826–1995). One strand of human geography at Harvard was the existence of the University Extension Program in the 1910s, which allowed non-Harvard students to earn credits toward an associate degree in arts (Harvard University 1826–1995). These human geography courses featured the first woman to teach geography at Harvard: Elizabeth Fisher (Harvard University 1826–1995). Soon thereafter, Harvard graduated its first woman PhD in geography (Wright and Koch 2009). These developments parallel the elevation of the field of human geography in the United States, then centered at institutions such as the University of Chicago (Whittlesey's alma mater and employer where he was on faculty when Harvard's Lowell administration recruited him).

President A. Lawrence Lowell's administration sought to start a human geography program as early as 1926 and succeeded in 1928. Martin (1980, 69) argues that administrative officers considered the establishment of a separate Department of Geography at Harvard in 1926. Bowman (1948) corroborates this in his own notes from the Board of Overseers meeting on October 11, 1948, as discussed in chapter 5. On October 24, 1927, FAS Dean Moore instead courted former visiting professor Raoul Blanchard of France's Université de Grenoble to head the program. The administration also wanted an assistant professor to work with and eventually replace

Blanchard in the program, and for this position, they recruited Derwent Whittlesey, evident in a letter from Dean Moore to Whittlesey in 1928:

> We have desired for some time to develop our instruction in Geography along the lines of population and economics. For that purpose we have invited Professor Blanchard to join us . . . but I think it probable that after a few years he may find it difficult to be absent from France. . . . We are asking you to come . . . [that] you can be asked to take charge of this work when he withdraws.[3]

Whittlesey, who worked at the University of Chicago at the time, responded to the dean's initial offer by enquiring about the future of geography at Harvard.[4] In his correspondence with Whittlesey prior to his arrival, Professor Kirtley Mather also confirmed the Department of Geology's support for human geography. Moore added that the program would have free reign to expand so long as Whittlesey "shows himself the man to take that responsibility."[5] It was on this basis that Whittlesey accepted Harvard's offer on April 27, 1928. As shown in chapter 2, Whittlesey's ability to build human geography at Harvard was, from the start, premised on his masculinity. Blanchard and Whittlesey thereby headed the new instruction in human geography beginning in September of 1928, with Blanchard departing a few years later.

The department organized geography into three types of courses by 1929: systematic (later: topical), regional, and physical (Harvard University 1826–1995). As figure 4.2 shows, courses in human geography went from constituting 0 percent of student enrollment in geography in 1927 to 79 percent by 1947 (Harvard University 1826–1995). Geography at the college attained its zenith under Whittlesey's leadership after World War II, growing from 102 students enrolled in seven courses taught by five personnel in 1927 to 634 students enrolled in twenty-four courses taught by twelve personnel in 1947 (Harvard University 1826–1995).

Geographers experienced important achievements and setbacks during the 1930s. In the fall of 1930, the department and the Lowell administration approved the hiring of Harold Kemp.[6] Next, the FAS promoted Whittlesey to associate professor in 1931.[7] Though Blanchard was nominally the head of the program, he was present only during fall semesters from 1929 to 1933 and completely absent in 1934, never to return.[8] In his absence, Whittlesey stated in correspondence that he and Kemp did the vast majority of the program's work in difficult circumstances. As evidenced

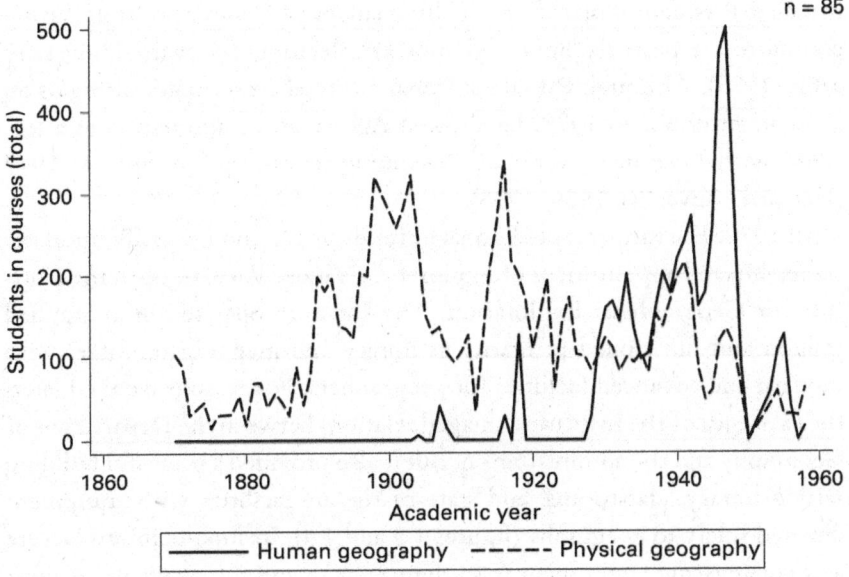

Figure 4.2
Frequency of student enrollments and classes in geography by subfield, 1870 to 1956.
Source: Harvard University, Harvard/Radcliffe Annual Reports (1871–1959).

in chapter 2, these circumstances included lack of funds and personnel to continue research and teaching. Even as the new program was deprived of resources, it blossomed under Whittlesey's leadership.

THE GEOGRAPHY PROGRAM'S EXPANSION IN THE 1930S

While narratives about the end of geography at Harvard often assume its decline in the time leading to its closure, we provide empirical evidence that disproves the assertion that geography was in decline in the 1930s and 1940s. Instead, we demonstrate the opposite to be the case. Geography's expansion continued throughout the 1930s. Between 1927 and 1940, the program increased from 102 to 418 enrolled students, seven to twenty-nine courses, and five to fourteen lecturing personnel (Harvard University 1826–1995). Whittlesey and Kemp expanded training in topical geography, making the university the leading campus for study of political geography in the United States under the leadership of a leading political geographer of the time (Morris 1962, 166; 241). While Dean Birkoff and newly installed President Conant eventually challenged Kemp's position, hiring committees repeatedly approved him as a lecturer (Harvard University 1826–1995). Although President Conant vetoed Ackerman's hiring as an assistant professor in 1939, he allowed Ackerman's continuation as a lecturer, with Ackerman eventually becoming tenure-track faculty in 1947 (Harvard University 1826–1995).

In 1932, Harvard accepted donated funds, made conditionally upon the leadership and appointment of explorer Alexander Rice, to open the Institute for Geographical Exploration. The Institute operated in an applied fashion and functioned primarily as library and mapping resource, with modern and advanced facilities for geographers. For reasons detailed later, the existence of the Institute damaged relations between the Department of Geography and the administration. But it also provided a beautiful building with a library, classrooms, and state-of-the-art facilities with equipment devoted solely to geography (figures 4.3 and 4.4). Its unique infrastructure and cutting-edge equipment (e.g., figure 4.6) were not available to most geography programs in the United States at this time.

Figure 4.3
The Geography Building, 1930s. Courtesy of Harvard University Archives, Harvard University Library.

Figure 4.4
The Map Library in the Geography Building, 1930s. Courtesy of Harvard University Archives, Harvard University Library.

The Institute's purpose was to train professional geographers in skills necessary for exploration, such as surveying, aerial photogrammetry, and cartography. Alexander Hamilton Rice founded it using funds provided by Eleanor Widener Rice, noted philanthropist and his spouse. According to the original agreement signed between Hamilton Rice and Harvard, the Institute would become part of the university so long as he funded a building to house its operations.[9] The "Geography Building" would revert to Harvard upon Rice's death, which would continue to fund its operation for another three years. The university would, in exchange, grant Hamilton a professorship—one which lasted until the Institute's closure in 1951.[10] The Institute thus created a second center for geography at Harvard, one which prominent geographers, like Isaiah Bowman, saw as depending solely on wealth for its existence and prestige, as we will explore further in the next chapter.[11]

Conant likely shared Bowman's antagonism toward the Institute, especially since it stood as a clear example of patronage through the Boston
elite, which he sought to eliminate (Hershberg 1993, 76; discussed further
in chapter 6). Geographers in the department also opposed the Institute
and actively contributed to prevention of its growth to protect what they
understood as scarce resources for themselves. A department committee
that included Whittlesey, for example, repeatedly voted against promotion of the Institute's staff, with Chair and Professor of Geology Donald
McLaughlin confidentially arguing that Institute staff did not deserve promotion and would inhibit other geographers from being hired.[12] While
granting geography additional facilities, instructors, and courses, the
Institute also generated opposition to the discipline in the administration,
thereby furthering uncertainty about the administration granting geography even more resources.[13]

The Department of Geology and Geography had presented initial plans
for human geography in its "Memorandum Concerning Plan for Expansion
of Work in Geology and Mining" on November 27, 1928 (Department of
Geology and Geography 1928). This memorandum outlined priorities for
organizing and allocating funding, space, and personnel to seven disciplinary fields, ranking geography as least important (Department of Geology
and Geography 1928, 7). This placement occurred because the department
expected geography to receive its own department:

> Geography is placed last on the list of units, not because its importance and needs
> are not appreciated, but because it is believed that with future growth it will
> probably become too large to be adequately accommodated within the present
> plan. As now, its needs could be partially met by continuing to share the facilities
> for general geology, but in the future it might be more effectively established as a
> separate organization, since its expansion is likely to be away from its contacts in
> geology. (Department of Geology and Geography 1928, 14)

Geography competed with and usually lost to geology for resources.
One tension owed to the perceived distancing of human geography from
geology, which Harvard's 1938–1939 Taylor Committee reinforced by
placing geography in programming for the social sciences—despite it being
organized in the natural sciences (Harvard University 1826–1995). Whittlesey argued that this difference caused geologists to fail to respect geography: "Very few geologists think geography worthwhile and that nearly

all suppose that they could easily do better at it than we do if they chose to take the trouble."[14]

Despite these tensions, Department Chair and geologist Donald McLaughlin hired Edward Ackerman as full-time teaching staff in April of 1939, until a more permanent position opened for the latter in the early 1940s. In a letter to Dean George Birkoff, McLaughlin noted that:

> At a meeting of the senior members of the department the other day, the appointment of Mr. Ackerman and the prospects for his advancement were discussed at length. . . . We were all emphatically of the opinion that instruction in geography and closely related subjects should be continued on the scale on which it was organized nine years ago. . . . We are all enthusiastic about Ackerman and believe that he is the type of young scientist we should endeavor to hold at Harvard.[15]

Ackerman would go on to become one of the central teaching staff in the Geography Program in the 1940s.

Geography attained its zenith at Harvard in the 1940s. The program made substantial progress toward becoming a department, enrolling more students, personnel, and resources. During this time, undergraduate enrollments and requests for graduate student research supervision grew more demanding, filling the lecture hall in the Geography Building, pictured in figure 4.5. Staff working in the Geography Program used that building's advanced technological facilities, as shown in figure 4.6, to conduct their research and teach.

During this time, Whittlesey himself attained repeated national recognition for his scholarship in political geography and for national leadership in the discipline. As noted in chapter 2, he served as the president of the Association of American Geographers for a year and the editor of the Association's *Annals of the Association of American Geographers* for eleven years.

The problems that geography faced at Harvard in the 1930s persisted, however, and the administration quietly began to unravel geography in the 1940s, as it grew, culminating in Conant's explicit declaration of a policy to "let the department die" in 1948.

In wartime, with close ties and pathways between geography and the US military, the importance, independence, and success of geography rapidly expanded in the 1940s (Barnes and Farish 2006). The department hired new staff, enrolled graduate students, and developed courses "because of their

Figure 4.5
Lecture theater in the Geography Building, pictured here in the 1930s and much the
same today. The building now houses the Department of East Asian Languages and
Civilizations and the Harvard-Yenching Library. Courtesy of Harvard University
Archives, Harvard University Library.

direct application to defense and war training" and as "knowledge necessary to win the War" (Harvard University 1826–1995). During World
War II, the program trained military officers and students in geographical
knowledge and state ideologies. Examples of the latter included promotion
of "hemispheric solidarity" through a "Human Geography of Latin America" course and offering "The Geography of the Soviet Union and British
Empire" (Harvard University 1826–1995). Whittlesey stated that this "war
work" employed all graduate students in geography from 1942 to 1945,
their number increasing from a few to sixteen by that time. The Committee
on War Service Credit required geography for its Foreign Area and Language Program, used to train administrators of occupied areas at the end
of the war (Harvard University 1826–1995). The federal government also
enacted the GI Bill in 1945, thereby funding over 9,000 veterans to study
at Harvard each year (Harvard University 1826–1995). Whittlesey wrote

Figure 4.6
Photogrammetric equipment in the Geography Building, pictured here in the 1930s.
Courtesy of Harvard University Archives, Harvard University Library.

that these changes, combined with chronic understaffing, created too large a workload for geography. He nonetheless was confident geography would continue to expand after the war.[16]

As the Geography Program became the prospective Department of Geography, its faculty hired new staff and expanded course offerings. The department and administration confirmed Whittlesey's position and accomplishments by granting him tenure and promotion in 1943 (Harvard University 1826–1995). This growth of the discipline at Harvard mirrored its growth nationally. Regarding his own promotion, Whittlesey wrote to FAS Dean Buck, noting gratification that geography had, at last, taken at Harvard a place reflecting its position in the world at large. Whittlesey's election as president of the Association of American Geographers in 1944 further reinforced this stature. The department and administration approved hires of Edward Ullman and Edward Ackerman as assistant professors in geography in 1945 and 1947, respectively (Harvard University 1826–1995). Human Geography thus reached its peak of four tenure-track and tenured professors in 1947: Whittlesey, Bryan (physical geography), Ullman (geography/urban planning), and Ackerman—in addition to lecturers and assistants on staff.

The Geography Program had long sought Ackerman's promotion from lecturer to professor, and Ackerman decided to remain at Harvard despite receiving eight jobs offers, including two chairships of large departments. To stabilize Ackerman's employment at Harvard, and without his knowledge, Whittlesey wrote to Dean Paul Buck in 1947, stating, "He is peculiarly loyal to Harvard. His refusal of many offers, tempting as to rank, salary, and opportunity, indicates that he hopes to stay here."[17] In the end, the program was determined to keep Ackerman at Harvard as a foremost representative of a new generation of geographers. Writing of Ackerman at this time, Kemp gave a glowing description: "He has travelled widely through the United States, won a 'major letter' as an undergraduate athlete, is a person of unusually wide interests both in and out of geography, a gentleman and a modest, unassuming person, in spite of an extraordinary success and a multitude of admirers."[18]

Harvard's administration planned to grant geography its own department. The Division of Geological Sciences, which included the Department of Geology and Geography, explicitly asked for the creation of a Department of Geography in a detailed report to FAS Dean Buck during this period of expansion, on July 1, 1944 (Division of Geological Sciences 1944). The "Geography at Harvard" report outlined how and where the Department of Geography would be created. Its plan included incorporation of Rice's Institute into the department, consolidation of all instruction and research into the Geography Building, and increasing staff to three full professors, three associate professors, one assistant professor, six instructors and four teaching fellows. The Institute would thereby become controlled and funded by Harvard, a plan that directly counters the notion that the Institute represented the undoing of geography. The Division of Geological Sciences proposed an annual budget of 112,800 USD (1944) per year (approximately 1.94 million USD in 2024), 66.5 percent of which would be covered by expected course fees and the rest through outside grants.

The department's research program proved as important as its material infrastructure. The division slated geography to focus on topical human geography and use fieldwork and statistical analysis as major methodologies. It would program courses in the social sciences and work closely with Anthropology, Sociology, Economics, Government, History, and Planning to achieve these ends. The analytical and material split with geology would likely correspond with the future Department of Geology's move to a new building on the "west side of Oxford Street" (Harvard University 1826–1995). This move corresponded with Whittlesey's belief that "without exception, geography has thrived in American universities only when it has achieved an independent status."[19] Geography's expansion and the emergence of institutional plans and resources to support this expansion emerged from Whittlesey's early plans, his tireless work with colleagues, his successful research program and national recognition, and successful enrollment and instruction of students.

THE GEOGRAPHY PROGRAM'S PROBLEMS: HOMOPHOBIA, MARXISM, AND THE "LET GEOGRAPHY DIE" POLICY

The Board of Overseers elected Professor James B. Conant president of Harvard in 1933 (Hershberg 1993, 65). As an outsider with liberal, meritocratic values, Conant's election was politically contentious among Boston's elite, conservative families, reducing his ability to solicit their resources (Hershberg 1993, 76). Conant reinforced this antagonism through his practice of acting without consent of the board, which largely represented the political interests of the aristocracy. As we explore next and more fully in chapter 6, these and other factors contributed to Conant's opposition to geography.

Additionally, Harvard's Taylor Committee led to the adoption of the Graustein formula and the "up-or-out" rule (Harvard University 1826–1995). The former prescribed strict quotas on the number of tenured positions available to each department; the latter meant that Harvard would terminate all faculty who committees rejected for promotion. As a result of these adoptions, geography and geology actively competed within the same quotas, and any failure to promote geographers would result in their removal. This created conditions where adding new positions in geography would likely be understood to happen at geology's expense, as we explore further in chapter 5.[20] This was a haunting precursor to the subsequent deterioration of many geography departments in North America, which featured divisions between physical and human geographers over a focus on their simultaneous and mutually beneficial strengthening.

Despite geography's continued expansion, problems of the 1930s persisted and deepened in the 1940s. In contrast with the report's plan for departmental status and resources, the administration began to quietly reduce geography's resources as the war ended. FAS Dean Buck, whose authority was enhanced by Conant's absence due to his work at the Manhattan Project, represented the administration's perspective:

> This demand [for geography] has largely disappeared. A glance at the course enrollment in Geography courses during the three terms of this year will reveal relatively small enrollments. With the expectations of further shrinkage in total enrollment in the next year, and in view of our great deficit, I cannot consent to the request for an additional appointment.[21]

Instead of falling, though, geography's total enrollment would increase from 317 to 634 students from 1945 to 1947, the two years after Buck made this prediction (Harvard University 1826–1995). Despite this rapid growth, the administration further reduced the program's resources by withdrawing tutorials by 1947.[22] It also removed geography as a field of concentration (known on most campuses as majors) and actively challenged the promotions of Ackerman and Edward Ullman and the hiring of teaching fellow Richard Logan in 1947 (Harvard University 1826–1995).[23] By threatening to reduce the faculty just as Kemp was retiring, the administration threatened the entire program. As discussed in chapter 2, Whittlesey argued that Ackerman's potential loss would foment a wider inability to recruit any further personnel. With Kemp's retirement, this loss of ability to source new faculty would bring research and teaching to a halt. Writing to geographer Richard Hartshorne at the University of Wisconsin, Whittlesey stated:

> You may guess, but in any event you ought to know, that if he [Ackerman] stays here geography will be a going concern, whereas, if he does not, the whole subject will fold up. This will not come all at once, because I expect to be on the job for 10 more years, but there will be no work other than my own if, and when, Dr. Rice withdraws his subvention. . . . I tremble to think what the unfortunate effects may be of having geography drop out of the picture at Harvard after 20 years. I never realized how the university world crutinizes [sic] every act of Harvard, until somebody calls my attention to it.[24]

President Conant showed little interest in raising funds for geography, a position well-established in the archives: "[We will enhance geography] when we have any funds to expand our work in this field. I am sorry to say that at present we are not contemplating any such development since, as you probably know, this is no time to indulge in new commitments for any university."[25] G. H. Chase, Conant's secretary, noted that this continued to be the case by the end of the 1930s: "Mr. Conant thought no expansion was possible unless new funds specifically for geography were forthcoming. I gathered the problem was somewhat involved with the future of the Geographical Institute [sic] about which there seems to be little certainty."[26]

President Conant perpetuated geography's inability to receive adequate funding. While the Great Depression had ended and the university was better resourced than ever, Conant opposed what he viewed as government intrusion into higher-level education (Hershberg 1993, 391; chapter 6). A chemist by training, Conant was generally skeptical of social science, except for its use in promoting nation-state interests. His perceived association of social scientists with "radical" politics, especially Marxism, also drove his opposition (Conant 1948, 199). These commitments limited Conant's support to cases where disciplines sustained what he believed to be core American values: "If coupled with practical experience and infused with a zeal to move American society along its historic road, [the social sciences] may be particularly effective" (Hershberg 1993, 36). Since the administration and federal government agencies considered suspect, if not seditious, any academic research or instruction regarding the Soviet Union, communism, or the "third world," political geography was an especially controversial subject (Diamond 1992). Its intellectual and colloquial connection with German geopolitics caused additional damage to its reputation.[27] As political geographers, Whittlesey and Bowman tried to deflect these antagonisms by separating political geography from geopolitics and conforming to contemporary scientific norms.[28]

As Smith (1987) observed, and as evidenced in chapters 2 and 3, homophobia also contributed to geography's difficulties at Harvard. The college had an established record of institutional homophobia since at least the nineteenth century, evidenced in incidents such as the 1920 Secret Court, termination and disappearance of Arthur Kingsley Porter, and the suicide of F. O. Matthiessen in 1950 (Wright 2006). These incidents transpired during Whittlesey, Kemp, and Ackerman's time on campus and showed that the administration's discrimination went beyond simple reprimand or even termination but involved active outing of allegedly queer people, thereby potentially causing them to lose their jobs, families, homes, and even their lives. Diamond (1992) additionally demonstrated how the persecution of leftist scholars connected to and intensified attacks on queer academics in the 1940 and 1950s. Rather than openly out and terminate suspected queer staff, the administration would tend instead to prevent their promotion, thus quietly leading to termination due to the "up-or-out" rule (Diamond 1992, Wright 2006).

Wright (2006) also shows how character attacks were commonly used to defame alleged queer people. Given that Bowman and Conant made such statements, as we will show, about Whittlesey, it was likely that Conant's homophobic opposition to geography at Harvard was both personal and institutional. We will explore Bowman's and Conant's perspectives further in chapters 5 and 6.

Geography was at a critical juncture at Harvard by the beginning of 1947.[29] Kemp was retiring, and committees had been established to assess promotion for the program's two assistant professors. Whittlesey remained hopeful that a department would still be created despite the administration's opposition. He focused his efforts on Ackerman's case, framing Ackerman as Kemp's successor and necessary for a department to "be able to go forward."[30]

Conant's administration initiated a policy to "let the department die," which eventually cemented the decline of geography at Harvard. Although this policy first appears in the archives in 1948, Conant clearly pursued this goal earlier, particularly by intervening in the promotions of instructor Logan and assistant professors Ullman and Ackerman in 1947.

Conflict over Ackerman's promotion would become the central flashpoint of events surrounding Conant's policy. Although an internal committee voted to recommend Ackerman's promotion in May of 1947, President Conant later vetoed this decision by August, thereby forcing it to be reviewed by a committee formed of experts external to the division.[31] President Conant briefed the committee on November 20, negatively qualifying questions related to Ackerman's case by stating that the prospective position would reduce resources among geology and regional studies and that it would have "no prospect of new funds" (see chapter 5 for more).

The external ad hoc committee nonetheless recommended Ackerman's promotion by December 9, 1947. Instead, President Conant vetoed Ackerman's promotion, terminating him. He explained his reasoning for this decision in a confidential letter to now Provost Buck on "whether Harvard University should proceed with geography or let the Department die":

> It seems to be that with all the demands on our funds we might as well make the policy decision of saying, "This is one of the things . . . that Harvard is not going to have." . . . I have never been able to see that there was a real need for a university Department of Geography or that the subject was a science in any proper use of the word.[32]

With this decision, the birth and death of organized geography instruction at Harvard were both tied to Whittlesey's employment. This outcome dealt a severe blow to the Geography Program and put plans for a department on hold, a hold that eventually became permanent.

DERWENT WHITTLESEY'S FIGHT TO SAVE GEOGRAPHY

While the Conant administration's policy on the Geography Program at the college slowly depleted its staff until Derwent Whittlesey was left alone, Whittlesey's archival records nonetheless reveal that he went to great lengths to preserve the program. His primary means to challenge Conant's policy was to organize political support among different groups, such as scholars, students, policymakers, and military staff, and to raise sufficient funds to counter Conant's claims that there were simply no funds available for geography (further discussed in chapter 6). He wrote in 1951, for instance:

> Perhaps you have heard nothing about all this. It created so great a stir when the announcement was made . . . [that] they pressed for a committee of faculty to study the whole question of geography. This committee of eight or nine representatives of different departments spent one afternoon a week for an entire college year and reported that geography should be set up as a separate department. The report has not been acted upon because the president says he has no funds for the purpose. I interpret this to mean that he doesn't want to spend the funds he has for this purpose.[33]

Whittlesey was, therefore, fully aware of the empirical inconsistencies that exposed the difference between official institutional narratives and underlying reasons. He echoed this challenge in other letters, such as this one to Charles Leonard, a geography alumnus working in private industry:

> The administration had it in mind to eliminate geography from Harvard, but so much objection sprang up that a faculty committee investigated the question through one whole college year and reported that it should not only remain in the curriculum, but should be established as a separate department. At that point, the President stated that there was no money for such a development. It therefore looks as though we could have geography any time we can get enough funds to launch it. Where such funds are to come from, I haven't any idea. Can you see any possibility from your favored seat in Wall Street?[34]

The tenor and content of these letters also demonstrate how Whittlesey's exhaustion mounted over time, moving from hope for geography's imminent revival to eventual resignation over its loss. While each institutional review called to recreate the program, and even though Harvard's college admissions had nearly tripled in a decade (Harvard University 1826–1995), the Conant and subsequent Nathaniel Pusey administrations continually refused to act, leaving Whittlesey's Geography Program as "unfinished business" and one of Harvard's greatest unresolved problems during the 1950s and early 1960s (Morris 1962). The urgency and potential of geography's rebirth became especially important when the Institute for Geographical Exploration closed in 1951, due to Hamilton Rice's lack of funds, leaving a fully furnished, specially designed building for geographical instruction available for its use (Stamp 1952). Whittlesey wrote to former graduate advisee Richard Logan, "I doubt if anything can be done about the immediate future of geography at Harvard. The real decision was made in 1948 and then every effort to obtain consideration at the most influential levels has failed when brought before the President."[35]

Derwent Whittlesey veered between hope and despair with respect to geography's fate at Harvard. Given geography's broad expansion in the United States during the 1940s and 1950s (Morris 1962) and the subject's importance, he believed that geography's absence would be addressed in time: "From all this it seems quite clear that the recession of geography at Harvard is only temporary and that Harvard will again take its place along with every other large university in the country."[36]

To this end, Whittlesey solicited President Pusey in April of 1954 to reconsider geography's situation; however, Pusey's administration continued Conant's requirement for an unnecessarily large endowment to create a Department of Geography—funds that ultimately never materialized.[37] Derwent's resignation grew as the powerlessness of his situation increased, witnessing not only the loss of one of his career's major works but also its resounding damage to the field of political geography, of which Harvard was a leading center. Corresponding with a parent considering sending their child to Harvard's Geography Program in 1954, Derwent noted:

> I wish very much I might look forward to having your son in geography here. Unfortunately the only courses given in human geography are mine and I shall be on deck only three more years after this. There is not a sufficient body of courses to take one [sic] anybody in the graduate school and therefore I can only regretfully advise your boy to go elsewhere.[38]

Regarding the frequency with which he received such letters asking to study geography at Harvard, Whittlesey complained to the new FAS dean:

> This letter seems to me cogent evidence of Harvard's opportunity to develop geography. I might say scarcely a week passes without receipt of one or more letters from foreign countries asking about opportunities for graduate study in geography here.[39]

After two years of trying to work with the Pusey administration, Whittlesey resigned himself to its continuation of Conant's policy: "Unfortunately, geography is apparently about to die out at Harvard."[40] Whittlesey understood that geography would die with his imminent retirement. Little did he know at the time of writing—February 20, 1956—just how literally correct his statement, for Whittlesey himself died, along with geography, nine months later.

The policy wiped out the Geography Program's staff, funding, and even infrastructure and forced Derwent Whittlesey to continue alone into his planned retirement in 1957. Despite being Harvard's only remaining geographical staff, the AAG elected him as its first honorary president in 1955, where he gave a lecture and chaired a panel on political geography at its annual conference. Ultimately, and regardless of what happened to geography, Derwent—always professional—called his time at Harvard "a pleasant occasion from beginning to end."[41]

THE PERSONAL AND PROFESSIONAL CONSEQUENCES OF CONANT'S
POLICY AND THE END OF GEOGRAPHY, 1948 TO 1959

Conant's policy to let geography die, which began as early as 1948, continued until the program's full demise by 1959. We will further explore
the joint necessary conditions for this result in chapter 6. For the moment,
it suffices to say that these conditions reflected President Conant's and,
later, President Nathaniel Pusey's personal verdict that geography was
not a subject worthy of cultivation at Harvard. This verdict persisted
despite at least five institutional reviews on the discipline's status between
1947 and 1956, which we review in chapter 6 as afterlives of geography's
death.

As for Whittlesey, while he remained at Harvard, taking a sabbatical
in 1952, he became the only permanent geographer left on campus for
four more lonely years in which he found himself taxed with the work of
carrying on alone, listing courses that had to be cut as a result of Conant's
policy,[42] and conveying to a discipline of colleagues and prospective students across the country what had happened and that no further graduate
research would be possible (Harvard University 1826–1995).[43]

Emeline's friendship no doubt sustained Derwent during this time,
enabling him to survive the devastating losses at work and the strain on
his personal life, including the absence of his dear friend and queer family
member, Edward. In her correspondence, Emeline writes often about Derwent's work—his research, writing, travel, and contributions on campus
at Harvard. She also shares her own work experiences on the college campus. In February of 1945, as Derwent and Harold work feverishly to build
geography into a department, Emeline writes: "You did not say, Derwent,
whether geography has yet been set up as a separate department. You told
me last spring that something like that was in the wind."[44]

When news of the closure eventually arrives in its finality, Emeline
proves in turn to be a dear and compassionate friend whose love sustains
Derwent. Derwent discusses his distress, lack of sleep, and decision to leave
Cambridge for a while. Harold, meanwhile, travels to Wooster to spend
time with Emeline, retreating and recovering from these events at her
home. During this time, Emeline learns more about what has happened
and is concerned for Derwent. She also observes how miserable and self-
absorbed Harold is at this time, even though Harold visits with his then

romantic interest, Michael, before Michael's departure from Harold's life (which seems roughly to correspond with Edward's going).

Here, Emeline observes the situation of geography at Harvard:

> As to the situation in the department there, Harold said that he came here to get away from it so I did not press my questions; I gathered, however, that there was so much opposition to the doing away with geography that there might be the possibility of a reversal of decision. . . . The disconcerting thing about these changes in life is that they come with such complete suddenness and irrevocably alter the course of one's existence. . . It was a cruel and unwarranted blow and, as you say, does extreme violence to good sense.[45]

Emeline, having witnessed the interwoven personal and professional struggles that Derwent and Harold endured over the years, narrates the closure and its devastating impact on them. She writes the following in March of that year, a couple of months after the decisions detailed in our prologue have been taken: "I can hardly believe that Harvard would adopt such a short-sighted policy as to eliminate a science so increasingly important as geography. It just doesn't make sense. . . . I can understand how you seem to stand in the ruins of your world. . . . But the feeling that your work of 20 years has gone into the ashcan is nonsense. . . . I am just terribly distressed and disturbed. It is hard to know that you are so unhappy."[46]

Harold writes the following on the back of a letter that Derwent is sending to Charles Colby, a friendly colleague and correspondent since the 1920s: "Everything Whit says [about Harvard geography] is true. We are in a mess, just when it seemed that we had found our place in the sun. . . . What effect it will have on other universities to learn that Harvard has abandoned geography, I hate to think. I think it may be a serious setback all along the line."[47]

Emeline notes Harold's discontent and malaise:

> I was truly shocked at his constant carping and criticism of everything and everyone. After he had been here a couple of days, Edith said to me, "Your cousin seems so scornful of everything." . . . He used to be interested in the house and enjoy it's tranquility, give me some advice about the yard and show some interest in my life. Now he is so absorbed in his self-created misery that his surroundings are negligible. I see such a change in him. . . . The only things Harold seemed to enjoy were, the drives and the Ohio scenery and showing Ed's pictures.[48]

In contrast, McSweeney shares her own recent happiness with Whittlesey, her pleasure over having new clothes and social invitations to visit and dine with others. "I am having such a good time, Derwent, and life seems so wonderful to me. It pains me deeply to be with those who have much and yet think only of their lacks. I can't bear to have other people unhappy."[49]

CONCLUSION

Derwent Whittlesey continued research and graduate instruction until his sudden death near the end of 1956. The continuation of the policy to let geography die forced Whittlesey to reduce graduate instruction, repeatedly rejecting and referring prospective students to other schools.[50] Research and instruction at Harvard effectively ended when Whittlesey died, with the college offering no geography courses in 1957. By 1959, the Department of Geology removed "geography" from its name, rejected all graduate student applications in geography, and no longer offered graduate degrees (Harvard University 1826–1995). This completed the decline of geography at Harvard. Geography's afterlives, however, continued to haunt Harvard's campus into the present—something we will return to in our concluding chapters.

Ed Ackerman resigned his position in August of 1948, eventually going on to a highly successful career as a geographer and planner, working as the vice general director of the Tennessee Valley Authority and as director of the Carnegie Institute, as documented in his personal archives, which are housed in Laramie, Wyoming, at the American Heritage Center at the University of Wyoming. Ackerman remained in close contact through his correspondence with Whittlesey and Kemp, which can be found in Ackerman's archives. We share examples of this correspondence in chapter 8. We know that he likely returned to Cambridge after Whittlesey's death in 1956, and served as executor of Whittlesey's will, a fact we affirmed with a visit to the City of Cambridge archives—pictured in figure 4.7—where Ackerman "of Washington State" is listed.

In this chapter, we have explored the longer history of geography as a field of study at Harvard. Despite geographers' successes and achievements, including popularity among students, Conant's administration quietly started to end geography at the college, culminating in Conant's formal policy by January of 1948. This policy was possible only due to its continued,

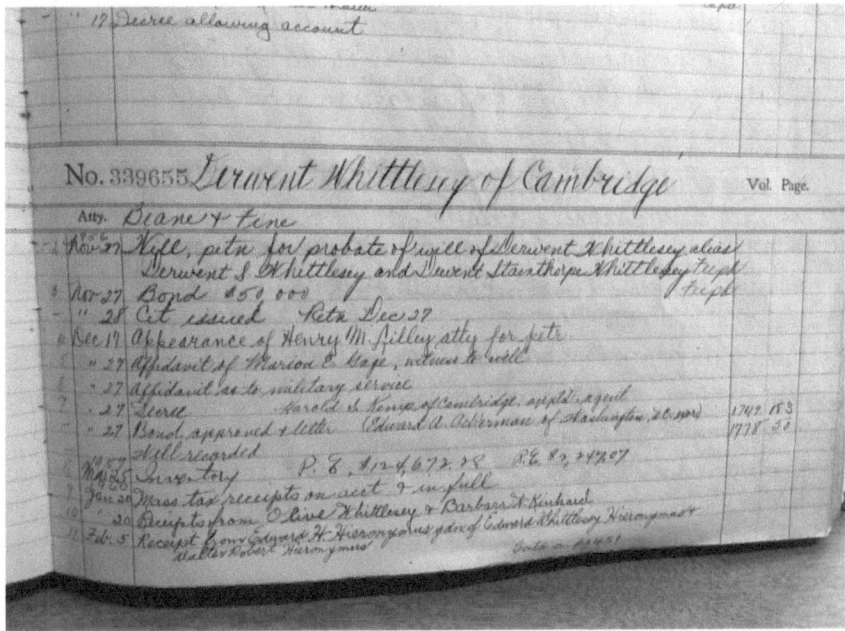

Figure 4.7
Photograph of entries related to Derwent's will, 1956, including listing of Edward Ackerman "of Washington State" as executor. Source: City of Cambridge Archives, Cambridge: Middlesex County Probate and Family Court.

active implementation by the Conant and subsequent Pusey administrations, which refused to alter their views or allocate any new resources. It proceeded despite at least five institutional reviews calling for the Geography Program's revival between 1948 and 1958, which we further explore in chapter 6. Starved of human and material resources, Conant's policy directly tied Whittlesey's death to geography's fall at Harvard.

Our analysis of new archival materials showed that the program's story at the college was more complex and different than previously imagined; we, therefore, provided a more elaborate account to address existing gaps in knowledge, while contextualizing the situation of the book's key characters, especially Derwent Whittlesey. Geography was not only well established at Harvard but became a leading center for the discipline first in physical and then later human geography during the twentieth century. Although the program reached its zenith under Whittlesey's and Kemp's leadership by 1946, the Conant administration started to implement a policy to end it by 1947. This policy, which cut off and diverted all of geography's limited human and material resources, thus tied its fate to Whittlesey's life.

Contrary to Smith's assessment (1987), we show that the Geography Program's central problems were based not on a lack of effective leadership by Whittlesey or Kemp but, rather, the interplay of forces of homophobia, McCarthyism, and the politics of scientific legitimacy. As we argued in chapter 2, of particular consequence was the construction of Whittlesey's ability to head a Harvard program based on their masculinity, which institutional and individual homophobia challenged. We will also further document the stories and antagonisms of Isaiah Bowman and James B. Conant with respect to the college's Geography Program in chapters 5 and 6, respectively.

As observed in chapters 2 and 3, geography's fall at Harvard did not go unchallenged by its staff, especially Whittlesey. Because the program was effectively his creation and central to his life, Whittlesey's archives contain detailed information about his response and perspective on the events surrounding Conant's policy. His correspondence with close friend and confidante Emeline McSweeney particularly contain personal reflections and descriptions of these events. We add to these correspondence through analyzing five institutional reviews on geography at Harvard after 1947 in chapter 6, revealing new evidence of Whittlesey's desperate battle to keep a program alive with limited resources.

A TRAGEDY IN THREE ACTS: ISAIAH BOWMAN
AND HARVARD GEOGRAPHY

Everyone has secrets, and deep inside his archives at Johns Hopkins University, Isaiah Bowman left one such secret for future discovery: a note on his personal antagonism with Alexander Hamilton Rice and Harvard's Institute of Geographical Exploration. The letter was to be opened only after Rice's death (figure 5.1; Bowman 1937). Bowman wrote the note and attached newspaper articles about the passing of Eleanor Elkins Widener, who had donated funds to construct Harvard's Widener Library after losing her son and experiencing her own near death on the RMS *Titanic* in 1912. The college and Widener's spouse, Rice, used her wealth to build the Institute, completed by 1931 at 2 Divinity Avenue (Morris 1962). Ironically, not only would Rice ultimately outlive Bowman but restrictions Bowman placed on his own archives, limiting their access to older men of international repute, would leave the note to collect dust until Johns Hopkins University archives overturned them and allowed public access in the 1980s—paving the way for the detailed work by Neil Smith (1987; 2003) on Bowman's life.

In archival research, one can sometimes glimpse an idea of what sort of personality a deceased person wanted to project to the world based on their own records. As we methodically reviewed Bowman's archives at Johns Hopkins University, his desire to be remembered as a prominent, masculine, and accomplished figure struck us deeply: Bowman not only believed intensely in the power of his own narration, he wanted to *store* that narration to set the record straight on topics important to him, such as geography at Harvard University and the roles the Rices played within it. In "writing this note to place on record her [Eleanor Elkins Widener's] relations," Bowman therefore sought to inject himself, albeit discreetly, into the history of geography at Harvard. As it would turn out, Bowman's entanglements would only deepen.

But back to our story.

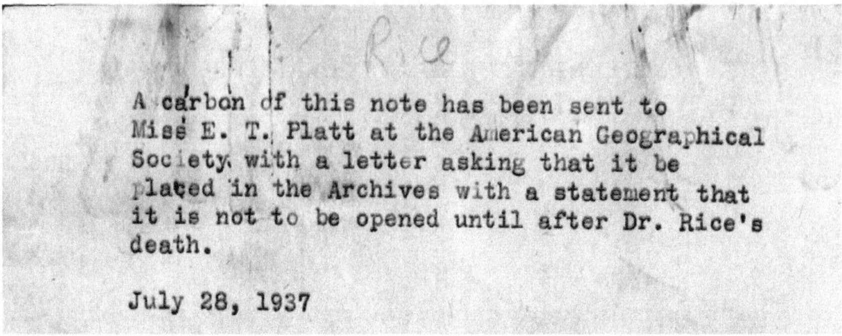

A carbon of this note has been sent to
Miss E. T. Platt at the American Geographical
Society with a letter asking that it be
placed in the Archives with a statement that
it is not to be opened until after Dr. Rice's
death.

July 28, 1937

Figure 5.1
Introduction to Isaiah Bowman's secret note on Eleanor Elkins Widener and Alexander
Hamilton Rice. Courtesy of University Archives, John Hopkins Sheridan Libraries.

Bowman's secret note documented the history of Eleanor and Hamilton
Rice in two key organizations advancing geography in the United States in
the mid-twentieth century, from his perspective: the American Geographical Society (AGS), which he directed from 1915 to 1935, and Harvard's
Geography Program. In the note, Bowman focused on criticizing the Rices
for their attempts to lead both organizations, as he saw it, by using their
wealth:

> At any rate, on one of these visits [in the late 1910s] she [Eleanor] said that she
> wanted "Hammy" [Hamilton] to be President of the [American Geographical]
> Society in place of Mr. Greenough. I told her the Presidency was a place one
> had to earn. . . . After the War and, as I recall it, about 1921 or 1922, I was asked
> to give a lecture at the War College of Newport . . . and after arriving received
> an invitation to stay at the Rice's. [Hamilton] Rice would not take No for an
> answer. . . . Both before leaving and afterward, Mrs. Rice referred to the "old
> fogey" whom we had as President (Mr. Greenough) and how very much better
> it would be if "Hammy" were made President. . . . I made no response to these
> suggestions except to counter them pleasantly with remarks of another charac-
> ter about the society and its work. But she was not interested in the Society's
> work: she was interested in her husband as president of the Society. (Bowman
> 1937, 1)

Bowman tried to control what he believed was the aristocratic privilege
of the Rices, supplemented by a further critique of the latter's forays into

defining what constituted true work in the field of geography, of which Bowman saw himself as the leading expert:

> I told Mr. Huntington [a council member of the AGS] that I would deplore any development which brought the Rices into closer connection with the Society. The reason for my statement lay in the open and increasing hostility of [Hamilton] Rice to me and the program of the Society. . . . He sneered at the work we were doing . . . and never failed in a general statement of policy or of program on my part to make a counter-statement, not so much opposing what I had said as insisting that he and other great geographers and especially foreign geographers knew what was "sane geography." (1–2)

According to Bowman, from there, his conflict with the Rices became acutely personal:

> Mrs. Rice stated categorically to Mr. Huntington that the Society could have all the money it wanted if they would make "Hammy" president. She would supply whatever was needed, but the Society would first have to meet a second condition which was laid down, and that was, as quoted by Huntington to the Council and to me, "You must first get rid of Mr. Bowman." In reporting it to the Council Mr. Huntington said ironically, "The question before us, gentlemen, is whether we are willing to sell Mr. Bowman for a million dollars, and I do not think we are!" There was a general laugh went up from the Council, and nothing further was said or done about the matter. (2)

Bowman then implied that following their failure to control the AGS, the Rices turned their attention to Harvard's Geography Program, where he continued to fight them in a valiant, though, as we shall see, ultimately futile effort:

> [Hamilton] Rice's next step was to offer to put up the institute at Harvard, which proposed in a letter to [President] Lowell and which Lowell accepted in a letter to Rice, to be called The Geographical Institute or the Institute of Geography. These two letters I have held in my hands. They were supplied to the committee of the Board of Overseers by President Lowell, for a question arose on the Board as to the desirability of entering upon such an arrangement. . . . [The committee chair] came to see me late one afternoon or early evening at my office at the A.G.S. They asked me what I thought the effect would be upon the geographers, if unfavorable, might not be harmful to Harvard. I agreed with them and we discussed reasons and alternative suggestions. . . . In

the correspondence, it was mentioned that he [Hamilton] was to be "Professor of Geography." (2)

Although unable to prevent the planned deal between the Rices and Harvard, Bowman used his influence to control the meaning of "geography" and lamented his perception of President Lowell's opposition to him on a personal level:

> After we had gone over the objections, [chair] Mallincrodt asked me if I would accept Rice's proposal if I were acting in Lowell's capacity as President of Harvard. I told him I thought I would, but that I thought I would try to get the titles changed to "Institute of Geographical Exploration" and "Professor of Geographical Exploration". [He] agreed, the matter was then thrashed out in the Board of Overseers, and the titles were adopted by the corporation and Overseers became the official titles as they stand today. I do not know whether Lowell ever learned from Mallincrodt or others that I had been consulted; but it is note-worthy that when my name came before the Board of Directors of the World Peace Foundation in Boston for election as a Director, Lowell spoke strongly against the proposal and when the vote was taken he voted in the negative. . . . At my first meeting of the Board I was called upon to make a report for a committee of which I was a member, and Lowell in a most disagreeable manner attacked the recommendations as if coming from me. On subsequent occasions . . . he ignored me on every occasion that it was possible to do so. (3)

Bowman went on to blame Rice for President Lowell's antagonism to himself and inferred that the entire enterprise was a folly serving Rice's personal ends:

> Of course his [President Lowell's] attitude might have been based upon what Rice told him directly about me. In spite of his attitude Lowell made a response one day to a question of mine which is worth putting in the record. We were the last to leave the Board room one day when he turned to ask me if I had seen Rice's new building and I said had not. . . . Then I added, "What does he propose to put in the building?" Lowell looked around the room as if to make sure that no one heard him and then said with lifted eyebrows and a knowing look, "Well, principally himself!" (3)

In his conclusion, Bowman issued a scathing attack on the existence of the Institute despite his claims of initial support, stating that Harvard University itself effectively sold Rice a professorship and that his work did not constitute true geographical research:

I heard from time to time of these increasing difficulties which Rice is having, his inability to secure the cooperation of other men in geography and climatology at Harvard, his absentee management, which means neglect of the Institute . . . the backing and filling of the Institute without any considered program of geographical research and the general foolishness of the work and publicity about it, which was out of tune with the rest of the University, and finally the circulation throughout Cambridge of the criticism that Lowell had sold a professorship to Rice in return for a building and an endowment . . . Evidently things had changed at Harvard and a professorship could now be bought, and the Institute of Geographical Exploration was a monument to the fact. (3–4)

Not only did Isaiah Bowman's secret note reveal aspects of his personality he wished to project, like prominence, masculinity, and accomplishment, they showed his commitment to being regarded as the leading figure of geography in the United States and his broader ability to influence Harvard University's Geography Program as such. Given Bowman's reputation, association, and connections with that program, it was perhaps no wonder that existing accounts (e.g., Smith 1987; Martin 1988) of the history of geography at Harvard placed him as central to its demise. In fact, he had tried to save the program.

What, then, can masculinity and attempts to control geography as a discipline, along with this historical record, say about the events at Harvard in this story? To examine this, we will first turn to a brief history of Bowman's life and then explore his intimate associations with Harvard's Geography Program, especially during the assessment of Edward Ackerman's bid for tenure in 1947 to 1948. After dissecting his personal responses to events related to President Conant's policy to let geography die with new evidence, we will conclude this chapter by returning to existing accounts of these events in order to reposition Bowman's role within the story.

ISAIAH BOWMAN: A BRIEF HISTORY

Isaiah Bowman's archival records are housed at the Milton S. Eisenhower Library at Johns Hopkins University.[1] Bowman possessed an intense, almost obsessive desire to be remembered, which became clear from the structure and contents of his records. An entire series of Bowman's archives, for instance, served biographical purposes, including thirty-six folders storing information ranging from cue cards describing his full honors; letters detailing formal awards, memberships, passports, and vacations; and hundreds of newspaper articles about himself. He carefully preserved and curated mostly autobiographical accounts of his life, sorted by date, which composed, by far, the largest portion of this archive. In discussing Bowman's life, we draw heavily from these sources to narrate his story through his own voice.

Isaiah Bowman (figure 5.2) was a Canadian-American geographer, geopolitician, and the fifth president of Johns Hopkins University (from 1935 to 1948).[2] He was born in Waterloo, Ontario, Canada—coincidentally where the two authors first met in 2012—on December 26, 1878, into a Mennonite family. Within two months of his birth, his parents moved as pioneers to Michigan, where he lived until 1902. During his formative years there, Bowman recalled his love of knowledge and its connection to masculinity:

> I cannot remember a time when the getting of knowledge was not an exciting business, and if there is one thing which runs through my entire life with complete continuity it is that I have always cared like blazes to know what was just ahead. . . . I had two older sisters who were studying more advanced subjects in school and talking about them at home. Their talk greatly excited me, especially what they said about the settlement of American and the Indian battles, massacres, and ways of life. This was no doubt merely an expression of the universal interest of boys in fights. My father and mother tried to keep me out-of-doors and away from books . . . [but they] drew out all my attachment.[3]

Describing his early schooling as "very poor," as his family endured economic hardship and he was one of eight children, Bowman nonetheless credited to it the formation of mental discipline: "It would take much more reflection than I can give at the moment to pass judgment on the general value of such discipline. I can only say that it appeared valuable to me."[4] Knowledge and education eventually became the basis of his career.

By 1895, Bowman passed a teachers' examination in Sinclair County, Michigan, and began to teach in three different school districts over the next four years.[5] In his records, he reflected on his "naturally" masculine-coded leadership qualities, as affected in a story of his self-organized military training during the Spanish-American War:

> At the beginning of my third year [teaching], in autumn of 1898, the effect of the Spanish[-American] War was felt everywhere, and I organized the young

Figure 5.2
Undated photograph of Isaiah Bowman. Courtesy of University Archives,
John Hopkins Sheridan Libraries.

men of my own school district and two adjoining districts into a private military company of 100 members, of which I was elected captain. . . . For two years we carried out military training, which included the manual of arms and simple maneuvres [sic] and field dispositions. . . . Looking back on these enterprises, I am struck by the uninhibited way in which I went about the whole business.[6]

Although Bowman finished four years of teaching in 1899, he was in a "deeply discouraged mood," having yet to begin college, never mind completing any customary preparatory school subjects.[7] During this time, Bowman recounted a story about what finally inspired him to pursue further studies:

One day my mother came out into the yard where I was chopping wood and the look of concern on her face was so marked that I asked her what the trouble was. She answered me by saying, "Son, I am worried about you. What are you going to make of your life?" I replied that I did not know, because I was nearly twenty-two years old and without any means for continuing my education. . . . Her comment was, "If I were a young man of your age and had your strength and interest in intellectual work, *I would go to college!*" . . . My mother and I were so sympathetic and understanding with respect to each other, that I took her advice and landed at a preparatory school.[8]

Bowman reported that, at this preparatory school, and subsequent studies at Michigan State Normal College (1899 to 1902; now Eastern Michigan University), that he had "the highest record made by any student who ever attended."[9] During this time, he decided to specialize in geography and worked with Professor Mark Jefferson, who had studied under William Davis at Harvard. According to Bowman, Mark would enable him to become a university lecturer if he, himself, would study under William at Harvard. He, therefore, went on to complete his studies in geography at Harvard University in 1905.[10] Bowman's experiences under William Davis and his approach to geography as a discipline began his ties to the field at Harvard and would heavily influence his thinking about geography.

Reporting that there were no positions in geography at this time, Bowman instead briefly pursued a career in the US Geological Survey, for which he claimed he came out "at the head of the list" for work in hydrology in 1905; however, Yale University soon approached him with an offer of a regular appointment. Bowman had earned his doctorate in geography from

Yale in 1909 and became an assistant professor of geography there from 1909 to 1915. Based on three subsequent field expeditions in South America during the early 1910s, the AGS in New York City selected Bowman to become its director in 1915—a position he accepted and administered until 1935.[11] Reflecting on his influence by this time, he stated that "in the course of twenty years [I] raised and spent about a million dollars. This was more free money than any other geographer in the world had at his disposal at that time."[12]

It was at this time that Bowman's forays into geopolitics began. While in New York, he became a member of the Council on Foreign Relations where, joined by other Harvard University alumni, he used geographical knowledge in leading others in the interests of American empire (Smith 2003):

> The most important event [at the Council on Foreign Relations] . . . was the invitation I received in late 1917 to join a small group of men working under the direction of Colonel House and authorized by President Wilson to prepare data and occasional memoranda for his use during the World War and subsequently at the Peace Conference at Paris. I became the executive officer of this organization which grew to contain 150 members. . . . President Wilson inspected the work at the House of the Society on October 12, 1918. When he sailed for the Peace Conference on December 13, 1918, a dozen of us were taken along to form the nucleus of an Intelligence Section of the American Commission to Negotiate Peace (as it was formally called) and of this I became the executive officer in addition to an appointment in the State Department as Chief Territorial Specialist.[13]

Based on this experience, Bowman wrote a book, *The New World: Problems in Political Geography* (1921), which he believed brought him into contact with wider geopolitical networks and thereby secured his subsequent appointments as director of the National Research Council in Washington, DC, and the Social Science Research Council in New York.

Around this time, Bowman claimed that Harvard contacted him to offer him a position as professor of geography: "In [the] [19]20's Dean Haskins said [the] administration [was] prepared to raise a million dollars if I would take a professorship in geography and be chairman of the Geology and Geography Department. I declined."[14] Eventually, Johns Hopkins University selected him to be its president in 1935.[15] This position, along with his future connections with Harvard President James B. Conant, would bring

Bowman back to Harvard when it appointed him as a member of the Board
of Overseers in the late 1940s.

President Roosevelt selected Isaiah Bowman in 1939 to head the secret
"M Project," which proposed mass "white resettlement" of European
Jews from 1942 to 1945, primarily in areas of Latin America and Africa.
According to Smith (2003) and archival records, Bowman was virulently
antisemitic and Roosevelt knew this. Roosevelt particularly believed that
Bowman would fail to cause political controversy by proposing any resettlement of Jewish refugees in the United States. Bowman actively refused
university positions at Johns Hopkins to Jewish people and instituted a
quota on their enrollment as students there in 1942 (246–247). The M Project files in Bowman's records documented previous experiences of colonial "success and failure," assessed the deployment of geography as a tool
to secure white settlement, and worked with the US Office of the Chief
of Engineers to create "engineering" reports for selected locations, such as
Brazil, Mexico, Angola, and Kenya, among many others. All of the settlement reports reached the conclusion that white settlement was possible,
and most added that it was desirable for a new, post–World War II world
order that would benefit the United States.

Bowman's geopolitical roles grew during World War II. During the conflict, he was a member of the Stettinius Mission of the Department of State to
London, chairman of the Territorial Committee of the Department of State,
vice chairman of the Post-War Problems Committee of the Department of
State, a member of the American delegation at the Dumbarton Oaks Conference of August of 1944, and a member of the American delegation at the
San Francisco Conference, where he helped found the United Nations (Smith
2003).[16] In these roles, and as discussed in detail by Smith (2003), Isaiah Bowman helped mold the American empire in the post–World War II period.

In 1948, Isaiah Bowman retired as president of Johns Hopkins University. Just before this time, as Harvard's Geography Program faced closure,
Harvard's FAS Dean Paul Buck and President James B. Conant chose
Bowman to intervene in an external committee on Edward Ackerman's tenure case. Amid these events, the college also nominated him to join its corporate Board of Overseers, where he became even more directly involved
with the Geography Program's fate. We present Isaiah's roles in these events
with new evidence later, as well as his response to them—which would
ultimately be cut short by his death on January 6, 1950 (Gladys 1951).

EDWARD ACKERMAN'S BID FOR TENURE AND PROMOTION, ACT 1:
ISAIAH BOWMAN ON THE EXTERNAL PROMOTION COMMITTEE

Bowman kept regular correspondence with Harvard President James B. Conant, as well as colleagues more generally, during the 1930s and 1940s. He categorized these letters in his archives into two separate folders: one on his general work and friendship with Conant, and one on his associations with Edward Ackerman's tenure and promotion case and work for Harvard's Board of Overseers. Their correspondence documented not only Bowman's more intimate connections with Conant's decision to let geography die but also the former's responses to that policy. Given the importance of Bowman's positionality in relation to these events, we present new evidence of his complex role in the Geography Program's demise in three parts, taking a brief interlude to examine Isaiah Bowman's relationship with Derwent Whittlesey.

Conant and Bowman had met by 1936, when the Governing Boards of Harvard University conferred an honorary degree of doctor of laws upon Bowman on June 18.[17] As presidents of major American research universities, the two men quickly formed a working relationship, where they consulted each other on varying issues, especially those related to academic hiring, research committees, and articles on World War II. In the course of this work, they became friends, discussing life difficulties, such as health problems, and stayed with each other overnight while in Cambridge or Baltimore. Though not the focus of this chapter, their friendship came into play during the episode of the Geography Program's demise at Harvard, where Bowman's closeness to Conant led him to argue for the *real* reasons behind the latter's policy, as we will see next. Far from being alien to each other, the relationship between these two men became a basis for yet more secrets behind geography's demise.

Based on Bowman's position as a preeminent American geographer, his connections with Harvard, and their friendship, James B. Conant first contacted him on Edward Ackerman's tenure case on September 25, 1947:

> I am venturing to trouble you about a Harvard matter. You may recall some time ago you were good enough to give us some assistance in connection with an *ad hoc* reviewing committee for considering a vacancy in the Department of Geography [sic]. We now have the problem of electing an associate professor. Dr. E. A. Ackerman, now an assistant professor, has been suggested. Such

suggestions are very carefully reviewed by a committee especially appointed for the purpose which includes men both from within the University and from without. Dean [Paul] Buck and I would very much like you to serve as a member of this committee.[18]

President Conant not only asked Bowman for his participation on this external review committee but also for help in picking other members to serve upon it; his influence, therefore, was potentially substantial. After accepting participation on Edward's external tenure case committee, Bowman stayed with Conant in Cambridge and met with the initial review committee on November 20, 1947, where he leveraged his friendship to push for his version of a prospective Department of Geography:

> Having said some things about the nature of geography that may have seemed too causal for the purpose in view, I am sending you a copy of a book which I wrote in 1934 as a member of the Commission on Social Studies that was financed by the Carnegie Corporation. The title of the book is *Geography in Relation to the Social Sciences*. . . . [The] comment on my phrase about geography being easy made me realize that my phrasing should have been a little more obvious. What I meant was that too many geographers giving courses in our universities are making geography easy. The facts of geography are generally interesting when presented in a mature way. . . . The earth-man relationship is vital to such a purpose. That is why I said last Thursday that I would not favor the establishment of a Department of "Human Geography."[19]

Bowman then proceeded to establish his vision for such a department at Harvard:

> The departments that have reduced or eliminated systematic work in physiography have suffered greatly. Their Ph.D. product is, for the most part, neither well-grounded in the physical principles that underline the phenomena of physiography and climatology, nor systematically trained in the principles of economics and political science, let us say. They seem to me to be suspended between earth and heaven and to offer neither good discipline nor particularly useful knowledge. What is needed, in my opinion, is a Department of Geography.[20]

After this meeting, President Conant contacted Bowman on December 4, 1947, to reply to specific questions regarding the outcome of Edward's promotion case:

> It was agreed when we broke up the meeting of the *ad hoc* committee on November 20 that I was to communicate with the members. I am therefore writing to you to ask two questions which I hope you will answer . . .

1. As a matter of policy would Harvard be well advised to make an appointment in Geography at the associate professor level, realizing in so doing
   the resources in Geology and resources in Regional Studies would be
   diminished by one-half place each; and at the same time with no prospect
   of new funds that would be necessary if any expansion of the department
   were made? In short, is it better for Harvard to attempt a partial solution
   of the Geography problem, or to let it be definitely the one-man department it is on a very limited basis?
2. If an appointment is made, is Ackerman the best man in the country
   available for this position?[21]

In two strikingly pointed letters, Bowman outlined his affirmative
responses to both questions. On the question of Harvard University's policy toward geography, he wrote:

> To answer the question of policy involves balancing (A) the disadvantages of
> establishing the new associate professorship in geography against (B) the failure
> to do so. It is my opinion that the latter outweigh the former and that the asso-
> ciate professorship should be established. . . . As to (1), it is true that Geology
> would suffer by the loss of a half a place. . . . I nevertheless feel that Geol-
> ogy at Harvard is in a strong position and that the loss of half a place would not
> be unduly crippling. . . . While there would be a difference in emphasis, the
> Regional [Studies] program would not suffer any net loss.
>
> As to (2), the matter of funds, is not a half of loaf better than a quarter
> of a loaf? While there is no present prospect of funds for expansion and little
> prospect in the near future, I should hate to think that the system of Harvard is
> frozen into such a rigid mold that a first-rate man cannot at least hope through
> first-rate endeavor ultimately to inspire the confidence that would gain for him
> the resources needed for the adequate development of the subject. Otherwise I
> should fear that teaching at Harvard would be rather discouraging.[22]

Bowman then continued by criticizing Harvard's policy on the subject in general, especially given its rising prominence as a field of study
worldwide:

> The disadvantages of geography's continuing to be inadequately represented at
> Harvard strike me as substantial. . . . It is not necessary today to strive to justify
> geography as a subject worthy of cultivation in the leading [American] univer-
> sities. The development of the field during the past twenty-five years or so in
> the government and in the colleges and universities seems to provide evidence
> of its cultural, educational, and practical values. . . . That geography *can* hold its

own with other subjects in universities with respect to both research and teaching hardly needs argument. Indeed, I am inclined to go further and claim that the failure to maintain adequate facilities for geographical research and teaching is a definite weakness in a university that aspires to maintain leadership. Geographical understanding is so sorely needed today that it would seem a positive disservice both to the students and to the members of faculty working in related fields *not* to maintain a competent department of geography.[23]

Meanwhile, on the question of Edward Ackerman's promotion for the position at hand, Bowman stated, "For the situation confronting the university now with respect to part-time work in area studies and part-time teaching at given levels, I think Ackerman is the best choice."[24] Most surprising, in his concluding remarks on the matter of geography at Harvard in his first letter, Bowman made a secret appeal to Conant to push for the creation of a Department of Geography:

> Although this lies beyond the scope of the committee, may I express a hope that some day a wholly independent department of geography can be set up at Harvard? The historical development that has kept geography as a sort of stepchild of geology in many universities has, in my opinion, hindered its growth. Geography needs geology but it also needs to stand on its own feet.[25]

In sum, Bowman not only supported the creation of an associate professor position in geography at Harvard University in 1947, he voted that the college should promote Edward to this position. He further made a wider call to President Conant to set up an independent Department of Geography, the absence of which Bowman claimed was a "definite weakness" in Harvard's educational leadership more broadly. As established in our previous chapter, therefore, Bowman was a critical part of the ultimately positive external committee's review of Edward's promotion case— which Conant effectively vetoed in his policy to let geography die during the events of 1948. Uncoincidentally, Bowman's role in the fallout from these events did not end here.

EDWARD ACKERMAN'S BID FOR TENURE AND PROMOTION, ACT 2:
THE BOARD OF OVERSEERS AND GEOGRAPHY AT HARVARD

In 1948, Harvard selected Isaiah Bowman to join its corporate Board of Overseers, one of the college's two governing boards. Ironically, the fate of

Harvard's Geography Program, and especially Edward Ackerman's promotion, would be the focus of his first meeting with the board on October 11, 1948. Bowman documented his critical interaction during that meeting with three sets of papers filed in two folders in his archives. The board's response, detailed in a report written by fellow members Lawrence Coolidge and Charles Wyzanski, became the central subject of the proceedings. We will discuss this report, as the first of five institutional reviews on geography at Harvard, further in our next chapter.

In likely preparation for this crucial meeting on his decision to let geography die, James Conant invited Isaiah Bowman to stay overnight with him at his house on 17 Quincy Street—an offer that was accepted.[26] Bowman's documents in his archives made clear that he was in regular correspondence with President Conant on the issue, both as his friend and fellow university president, before, during, and after the Board of Overseers' meeting. These interactions occurred not only over dinner on the night of October 10, but continued at a reception at Columbia University the next day, in addition to a subsequent meeting at the Brookhaven Laboratory Conference on October 13.[27] Based on Neil Smith's (1987) specific argument regarding Bowman's silence at the Overseers' meeting, we will work backward from a note he lodged on the subject from the Brookhaven Laboratory Conference on October 13 as new evidence from his notes summarizing his views on what happened at the meeting on October 11 (two days prior).

In this note, Bowman showed President Conant's apparent mockery of the situation in juxtaposition to his own view that geography should be expanded at Harvard:

> In the course of the day [of October 13] Conant and I conversed about the Board of Overseers meeting on October 11. He laughed heartily over the fact that on my first day at a Board meeting *geography*, of all subjects, should come up and the Board should show opposition to his action.
>
> "But," I said, "you must have noticed that I was silent, and guessed the reason." He replied, "I shall be grateful to my dying day for that silence. I think it was a remarkable piece of self-restraint, and I shall never forget it."[28]

That reason, as we will explore later, lay hidden in Bowman's notes on the meeting itself.

Amid this conversation, Bowman continued by attacking Hamilton Rice's work at the college: "Dodds came up and Conant repeated the story to him and I said to both that I did not see how anyone could respect geography who got his measure of the subject through dealings with Hamilton Rice!" Yet, at the same time, he used this opportunity to defend geography as a legitimate science while criticizing human geography:

> At dinner in the evening of October 13 at the Lab, Conant, Rabi (physicist, Columbia), and I had a three-cornered talk on science from the historical standpoint—the importance of study of great leaps of thought—Conant on Lavoisier and Boyle and I on Foucault's pendulum experiment and Eratosthenes first measurement of the size of the earth. . . . My purpose was to show Conant that geography had as good examples as chemistry—and as much claim to consideration as a science. The "human geographers" throw away this scientific heritage and step outside the great traditions of science when they limit their work to descriptive elements.[29]

So, even via his self-reported "silence," Bowman used his personal connection with James B. Conant to advance his call for Harvard's creation of a Department of Geography and defend geography's legitimacy as a science, as contrary to Conant's position. While Bowman's quotation on his silence, therefore, affirms Neil Smith's (1987) argument, our interpretation of that silence and its meaning simultaneously differs, as we shall see next.

But Isaiah Bowman was not silent at all on the issue of geography at Harvard during the Board of Overseer's meeting on October 11 in his archives. There, he left behind nine impassioned, meticulous pages of personal notes on the subject (figure 5.3). According to Bowman, the board meeting began at 11:00 a.m. in the Faculty Room at University Hall, where he was joined by the board's twenty other members, in addition to President Conant and his personal secretary, David Bailey.[30] Former Governor of Massachusetts and then US Senator Leverett Saltonstall chaired the meeting.

After reviewing the issue of new board appointments and other minor university matters, the meeting delved into the subject of geography at Harvard in earnest:

> The fun began with Coolidge's report on geography. It was thorough and pulled no punches. The issues were (1) need for geography in education, (2) specific needs of social sciences for geographical instruction, (3) legal questions that arose, and (4) mistakes of Conant in declaring his personal opinion of geography and then ordering the Provost to inform geology and geography faculty that the subject is to be abolished—that is, in determining a major policy

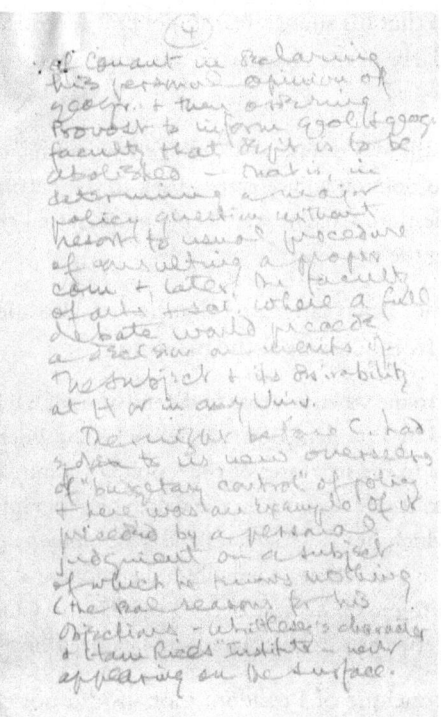

Figure 5.3
Introduction to Isaiah Bowman's notes on the Board of Overseers meeting.
Courtesy of University Archives, Johns Hopkins Sheridan Libraries.

question without resort to the usual procedure of consulting a proper committee and, later, the Faculty of Arts and Sciences (FAS), where a full debate would
precede a decision on the merits of the subject and its desirability at Harvard or
in any university.[31]

While allegedly remaining silent during the meeting, Bowman nonetheless used this opportunity in his archival record to criticize James B.
Conant's policy where he outlined the reasons for his silence:

The night before Conant had spoken to us new overseers of "budgetary control" of policy and here was an example of it preceded by a personal judgment
on a subject of which he knows nothing, the real reasons for his objections—
Whittlesey's character and Hamilton Rice's "Institute"—never appearing on the
surface.[32]

Bowman noted that his silence was in general accord with the rest of the overseers, particularly in the absence of any support for President Conant's actions:

> The room was still as a mouse as Coolidge reached his climax—presidential overreaching and opinion and reference back to the Faculty Arts and Sciences of the whole question. Conant was deeply embarrassed and evidently had no sympathy among the overseers.[33]

At the same time, Bowman stated that he was also silent due to the personal antagonism from Conant at the meeting:

> Conant referred to me when he arose to defend himself, saying he would "probably be criticized by President Bowman when he and I debate this matter later on." His answers to the strictures of the Coolidge committee report were dialectical and weak. It said nothing on the worth of geography. His letter to Provost Buck, in which he had characterized geography as not a science and as having no place in a university curriculum, he said was a "private letter"—but as Coolidge said to me on his way over to the Faculty Club for lunch, Conant could not call a letter private that stated a policy decision.[34]

But Bowman's critique of President Conant did not end here:

> Incidentally, the discussion made clear, and Coolidge's text was also explicit on the point, that the *ad hoc* committee was called to advise on Ackerman. What Conant did was to expand his decision beyond Ackerman's appointment to a final policy regarding geography. Nowhere did he call for expert and deliberate consideration of the latter question. Yet in the evening before, Conant had said that in faculty meetings he stimulated and even guided discussion but did not express an opinion or advocate adoption of his own views! "The president exceeded his authority," said Coolidge.[35]

He then said that the Board of Overseers moved to adopt the Coolidge committee's report and refer the entire question back for further review by the FAS. Bowman proceeded to spend the final three pages of his notes attacking President Conant's "extremely weak" response to this decision.

According to Isaiah Bowman, and as we will further demonstrate in chapter 6, Conant's response listed the following reasons to justify his policy. First, as previously mentioned in this chapter, the allocation of a professorship to Edward Ackerman would come at the alleged price of professorships in geology and regional studies. Second, there was no money for geography nor could any be located. Third, universities must specialize, and, therefore, some fields of study could not be accommodated. Fourth, his letter on his policy to let geography die was private. Fifth, the Department of Geology and Geography had waited too long in reacting to his decision to abolish geography at the college. And, last, geology would be unwilling to give further resources to geography. For each of these responses, Bowman laid out his own critique—in the archives—of why they were unfounded.

Ultimately, Isaiah Bowman's looming presence on Harvard's Board of Overseers meeting complicated President Conant's attempt to abolish geography as a field of study. Although he was *verbally* silent during the meeting, Bowman nevertheless supported the board's report that challenged Conant's overstep in authority regarding Ackerman's promotion. He also took detailed notes, where he attacked the president for his actions. In this way, Bowman navigated a situation where, as he would later write to Derwent on March 2, 1949, "I think matters can be worked out more quietly."[36]

Perhaps most importantly, the subject of Derwent Whittlesey himself and his "character" were central to President Conant's decision, as recorded in these handwritten notes by Bowman.

INTERLUDE: ISAIAH BOWMAN'S ASSOCIATION
WITH DERWENT WHITTLESEY

Amid the growth and eventual end of geography at Harvard, Isaiah Bowman kept regular correspondence with Derwent Whittlesey. In Bowman's archives, their association began by 1937, when Whittlesey asked Bowman to respond to a request for information on Harold Kemp, and continued until just before Bowman's death in 1950.

As discussed in chapter 2, during the 1930s, the FAS put Harold Kemp's position as lecturer in the Human Geography Program into question. On this basis, Derwent reached out to many prominent geographers in the United States to support Harold, including Isaiah Bowman.[37] Bowman ultimately assisted Whittlesey in this defense, and Kemp's hire came through without incident.[38] By this time, and as evidenced in these letters, Bowman and Whittlesey had an established collegiality and worked together to advance the interests of political geography as well as the discipline of geography as a whole.

In their correspondence, Bowman and Whittlesey frequently discussed issues involving geography in the United States, including its role in higher education and in military training during World War II. Based on their mutual interests and work, the two wrote to each other with a close and easy rapport. In January of 1941, for instance, Whittlesey made an exception of listening to talk radio to hear Bowman speak on *America's Town Meeting of the Air* regarding the Selective Service System.[39] The two also discussed and collaborated on publications related to political geography, such as Derwent's proposal to create the first *dedicated* journal to political geography—an idea which Bowman argued was ill-timed in 1945.[40]

Overall, Bowman believed that the joint efforts he was making with other geographers, such as Whittlesey, would culminate in substantially expanding the discipline, particularly with respect to the structure of a hegemonic world in the postwar period. In 1944, for instance, in anticipation of the end of the war itself, Whittlesey wrote to Bowman regarding the importance of geography in the postwar world prior to an upcoming joint meeting of the American Association of Geographers and the American Association for the Advancement of Science.[41] In response, Bowman said of this mutual work, "Certainly we have a great opportunity in geography; the greatest that we have ever had, to build its substantial elements into our educational system and indeed into our national thinking."[42]

In their final letters, Bowman and Whittlesey exchanged comments regarding the situation of geography at Harvard. In one letter, Whittlesey addressed Bowman on this issue as follows:

> After you left yesterday, I realized that you may not know the chronology of Human Geography here at Harvard. I was brought here to work with Raoul Blanchard in 1928, with the understanding that geography would be allowed to grow as fast as the demand warranted. This was either two or three years before Dr. Hamilton Rice made his proposal to President Lowell. From this it is clear that Harvard committed itself to geography with no reference whatever to money afterwards hoped from Dr. Rice. Since there is no one on the present administrative staff here who was active at the time, it is easy for the current incumbents to forget that the role of Dr. Rice was unrelated to the original plan of President Lowell.[43]

Evidently, while attempting to solicit Bowman's help in defending geography at Harvard, Whittlesey was unaware of the former's personal enmity toward Hamilton Rice and the Institute—a fact established by Bowman's response: "The sequence of events which involved Hamilton Rice were well known to me. One day I may tell you a part of the story as I knew it at the time."[44]

Overall, while not necessarily friends, Bowman and Whittlesey saw themselves as close colleagues working to advance their shared goals for the success of geography as a discipline. While the probable homophobic attack on "Whittlesey's character" in Bowman's Board of Overseers' notes was almost certainly reflective of James B. Conant's perspective on the issue of geography at Harvard, we cannot conclusively say that it was his own. That said, given Bowman's background and personal beliefs, as well as the pervasive homophobia existing at that time, we speculate that Bowman likely also accepted homophobic sentiments toward Whittlesey—which he may have concealed within his collegiality. On the whole, then, the two had a complex association reflected by Bowman's own sense of masculinity and worldviews, in addition to social forces at play in their joint contemporary context.

EDWARD ACKERMAN'S BID FOR TENURE AND PROMOTION, ACT 3: ISAIAH BOWMAN'S REACTION TO CONANT'S POLICY

James B. Conant's policy to let geography die at Harvard evoked a complex reaction from Bowman. As previously noted, Bowman made a plea— unusual for his character—to Conant to not only save Edward Ackerman's professorship but also propel an independent Department of Geography at Harvard. Furthermore, while Bowman remained verbally silent at the Board of Overseers meeting, he simultaneously wrote a scathing critique of Conant's actions within and beyond that meeting. Given Bowman's importance in our story thus far, and his centrality within previous research on geography at Harvard, we carefully examined various documents within his archives, especially his correspondence with FAS Dean Paul Buck and Professor Edward Ullman, to trace his ultimate response to the events at Harvard in 1947 and 1948.

As noted, prior to his appointment to the Board of Overseers in 1948, Bowman was also head of Ackerman's external review committee for his tenure case. Based on this position, Bowman collated and stored a series of letters and related documents on Edward's promotion, some of which expressed sentiments regarding the situation at Harvard. Writing to Paul Buck, for instance, he stated in 1948:

> From time to time I am in receipt of a letter from hither and yon to the effect that Harvard has dropped geography and why don't I do something about it. Let me say that my general reply is to the effect that I propose to mind my own business.
>
> But the number of such requests has aroused my interest. I was told that a statement on the subject was in the report on the Harvard Committee on Education. Unhappily there is no index to that report and I have been unable to find any relevant paragraph. . . . I would appreciate receiving a copy of any formal statement, provided that such a statement is available for distribution outside administrative circles at Harvard.[45]

This letter expressed, on the surface, Bowman's bifurcated reaction to President Conant's policy: nonintervention and tolerance, largely based on his position as a fellow university administrator and friend to Conant himself. As we have seen, sentiments along these lines conveyed haunting silences along the lines of masculine territorialities, with each character attempting to effectively demarcate their perceived self-ownership of the situation of geography at Harvard University—a theme to which we will return in our next chapter. From this angle, Bowman's *official* response to the cascading events leading to the demise of geography at Harvard implied his perception of a lack of authority to determine its fate. Publicly, therefore, he deferred to the Conant administration's actions.

The deeper level of Bowman's reaction to President Conant's policy, however, was private protest. In 1949, for instance, Edward Ullman afforded Bowman an opportunity to oppose geography's ongoing erasure at Harvard via a geography conference being held there:

> The New England Geographical Conference is holding its annual meeting at Harvard this year. We should like very much to have you talk to the organization on Saturday, May 7. This date is the nearest Saturday to the Overseers meeting on May 9, which I understand you expect to attend. We could change the meeting date to the preceding or following Saturday, if that would fit your plans better.
>
> I hope very much that you will accept, not only because the conference wants you, but frankly because it is an opportunity to do something toward reviving geography at Harvard. This meeting should provide an excellent platform for advocating geography in this critical area.[46]

In responding to Ullman's call for support, Bowman erroneously assumed that his proposed talk would necessarily be *on* the subject of geography at the college—and thus require a public rebuke of the Conant administration itself:

> What you say about geography at Harvard and a feature interview requires me to say that I believe the future of geography at Harvard to be secure when another round of discussion takes place. My relation to my fellow-members of the Board of Overseers and to President Conant would hardly permit me to give a feature interview as proposed. I think matters can be worked out more quietly. What I have said about myself does not apply to other speakers who are in a different relationship to the administration and it is altogether necessary to say in support of their position.[47]

While he ultimately declined to speak at this conference, Bowman scheduled a personal meeting with Edward Ullman about geography while in Cambridge in March.[48] Following this meeting, he sent Ullman materials to assist in the latter's work on the Committee on Geography at Harvard—the second of five institutional reviews we will examine further in the next chapter.[49] In his own handwriting on a typed letter, Bowman added, "You may also find some 'ammunition' in my article issued April, 1949, in the Geographical Journal (London)."[50] These stories reveal Bowman's personal complaints with the ongoing events at Harvard, along with his belief that the issue could and would be resolved through private channels that remained internal to the university itself.

CONCLUSION: ISAIAH BOWMAN AND GEOGRAPHY—AN EXTENSION OF NEIL SMITH AND GEOFFREY MARTIN

As this chapter showed, Isaiah Bowman had a complex connection to the history of geography at Harvard. In our interpretation of events and Bowman's role in them, although he certainly opposed some aspects of the burgeoning Human Geography Program, Bowman ultimately worked behind the scenes in the hope to forestall and even reverse geography's decline at the college. While it was true that he held substantial power to affect the discipline's fate, available evidence suggests that limits to this power existed in the face of President Conant's dogmatic attitudes and isolated actions—not to mention potential damage to their collegial association and friendship.

That Isaiah Bowman was a key figure in this history was no mystery for earlier researchers and previous accounts. Bowman's imposing presence emerged not only within such accounts but also in the mythology and urban legends of what had happened to geography. But unlike our story, particularly in the work of Neil Smith and Geoffrey Martin, we found Bowman to merely be important but not the *key* figure of geography's downfall.

Given the centrality of Bowman's role in these histories, we will now turn our attention to what our new archival materials and analyses thereof can imply about him and previous accounts. We do not seek to refute these accounts but, rather, to *extend* them by discussing how our data contend with Smith's and Martin's stories. We will then stage a brief historiographical intervention, one that explains how and why the *way* in which these histories of geography at Harvard overlooked the fundamental reality of Derwent Whittlesey's life and how this continues to haunt us, the college, and the discipline as a whole into the present.

In Neil Smith's published essay on the history of geography at Harvard in the *Annals of the Association of American Geographers*, he contended that Bowman's fundamental role in what happened to geography contributed to its final loss and subsequent erasure. Smith reached this conclusion for two major reasons. First, Derwent Whittlesey was, in his final analysis, too weak and ineffective to save the program. Second, and heavily based on a single interview with geographer Jean Gottman in 1982, Isaiah Bowman fundamentally opposed the program and, therefore, remained intentionally silent, failing to rise in its defense despite his influential position.

Given what we suspect was the controversial nature of Smith's piece during the review process, the *Annals* apparently (by announcement of its author) invited a subsequent commentary on the issue by historical geographer and official archivist of the association, Geoffrey Martin, in 1988 (Martin 1988). Though Martin agreed with Smith's basic claims, he devoted his commentary to an encompassing, hero-worshipping defense of Bowman—a scholar who he not only deeply admired but who was critical to his own research (e.g., Martin 1980). While our story so far has focused heavily on the importance of constructions and deployments of masculinity in the homophobic oppression Whittlesey, Kemp, and Ackerman faced, there ironically was a pervasive territoriality in Smith's and Martin's own conflict, whereby each of these men attempted to own the story by demarcating who and what mattered. This meant that masculinity was not only vital to this history but also to *the way in which it was told*.

We cannot do justice here to the casual homophobia that existed in these previous, important accounts of the history of geography. Without reflexivity on their positionality, both Smith and Martin discussed Whittlesey's and Kemp's character in terms coded for their sexuality. As was usual in the contemporary period, and as we have sadly seen throughout our story so far, these researchers manifested their own opposition through character defamation and erasure of Whittlesey and Kemp's relationship despite the overwhelming evidence of its existence.

But the problem became even more nefarious, insofar as homophobia itself modified both the events that transpired *in addition* to how the story was told. This came not only through direct homophobia in these pieces, which we will not quote here, but through indirect homophobia regarding which identities were present, as well as the bar to establish their existence. Effectively, had this been a straight relationship, we argue that no one would have challenged it, and, further, that the requirements to assert its validity would have been far lower. Instead, and as Smith and Martin repeatedly wrote, Whittlesey and Kemp were simply "good friends" and their association was a chief impediment to geography at Harvard.

That an apparent paradox emerged here was clear: on the one hand, Smith's and Martin's conflict profoundly relied upon the operation of structural forces, as embodied through particularized institutions and personalities; yet, on the other hand, they averred from the structural forces affecting *them* in their pursuits to claim the landscape of history of geography. Our goal is not to place blame or criticize Smith's and Martin's works, as, without them, we may have never arrived here ourselves; rather, we seek to understand how and why it was possible that a history could be written in this way.

It was no surprise that primarily focusing on structural, institutional forces would sublimate the hidden, obscured margins of the real life of Derwent Whittlesey in this story. Yet what perhaps was even more compelling than this was the way in which it contributed to his haunting us in the present. For, as an example, within the very same boxes and folders from which Smith and Martin quoted to establish their histories laid evidence that contradicted their analyses—be it Harold Kemp's love letter, Emeline McSweeney's correspondence, or even Isaiah Bowman's secret notes. If the Geography Program died like James B. Conant intended, then we must now acknowledge how Derwent's ghost continues to haunt us.

EVERYONE HAS SECRETS: CONANT'S CAMPAIGNS
AND CONTRADICTIONS

I still feel that Conant owes you (and me too) an explanation which has not yet been given.[1]
—Edward Ackerman, letter to Derwent Whittlesey, March 7, 1948

Everyone has their secrets, and it turns out that Harvard's President Conant did, too. So intense proved the burden of carrying these secrets, that President Conant—like his friend President Isaiah Bowman—felt the need to write to future "men of some importance." On February 7, 1951, as James B. Conant approached the end of his last term in office, he penned a secret letter to the future president of Harvard, handwritten in cursive in blue ink. The letter was sealed, with instructions that it be opened by the president fifty years hence, in the year 2000.

Dutifully, then President Drew Faust opened the letter as instructed in 2000, reading it to the public. The letter—shown in figure 6.1—opens, "To the President of Harvard University in 2000 My dear Sir" (an awkward and inauspicious opening, no doubt, for President Faust on the day that she read it aloud). After reflecting on a dinner to celebrate the 300th anniversary of the college at Dunster House, Conant wonders aloud whether the university will still exist when it comes time for the 350th Celebration.

By his second paragraph on the first handwritten page, Conant's foreboding turns to international events: "The international situation seems to have taken a turn for the worst." He continues, noting his own relief that "the nation" is finally now alerted to the "present danger:" "At all events, there are many who anticipate World War III within the decade and not a few who consider the destruction of our cities including Cambridge quite possible. Therefore," one can imagine him exhaling, "if you read this at least one set of prophets of gloom, at least, will have been proved wrong.

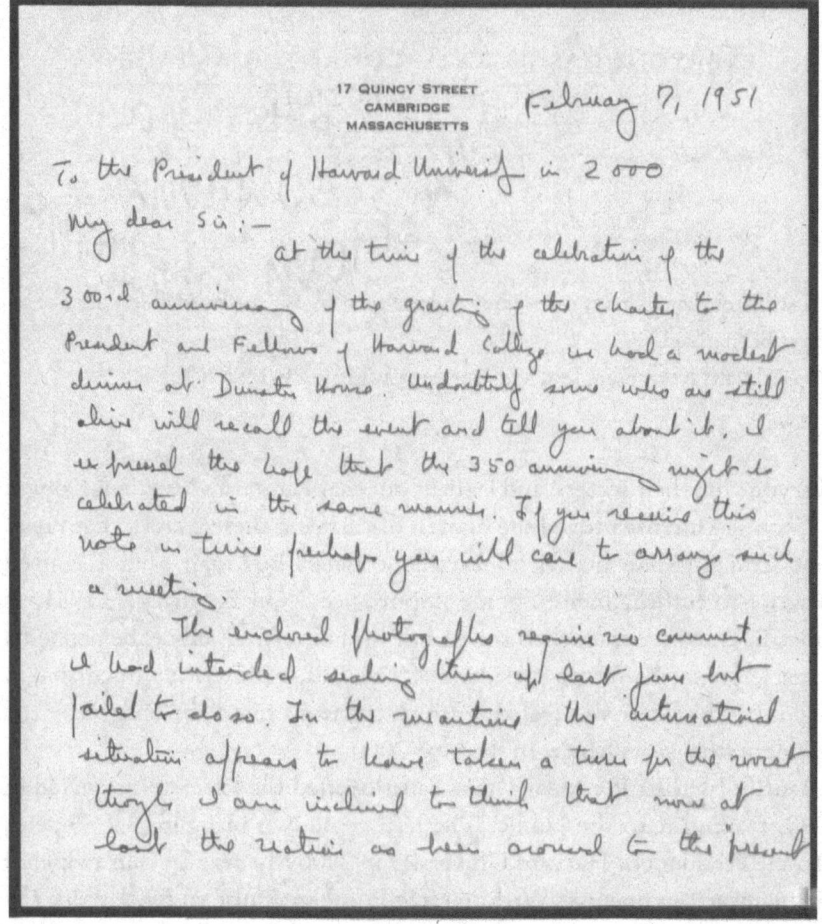

Figure 6.1
Photograph of President James B. Conant's letter to the future President of Harvard,
penned in 1951. Courtesy of Harvard University Archives, Harvard University Library.

We all wonder how the free world is going to get through the next fifty
years. You will know the answer."

Conant then reveals further thoughts on the vulnerability and prospects
of Harvard: "As you will see I am looking forward to the fact that you will
receive this note and be in charge of a more prosperous and significant institution than the one over which I have the honor to preside." He continues:
"How free it may be or rather how independent of the state is another

question. But that it will maintain the tradition of academic freedom, if tolerance for heresy I feel sure."

We find this question of freedom from the state both ironic and revealing, given Conant's own role in McCarthyism and the Red Scare, seeking to remove people with communist leanings from American university campuses during this period. Conant engaged lists of so-called reducators—educators with allegedly communist leadings who he committed to remove from their influence on American campuses.

We also find Conant's belief in the enduring nature of academic freedom deeply ironic, given his role in aggressively, and against substantial opposition, shutting down research and teaching in the field of geography at the institution of higher learning over which he presided.

This letter ultimately reveals more about James B. Conant himself—pictured in figure 6.2—than it does anything else: specifically, that he carried the weight of the world on his shoulders, doubting the continued existence of Harvard, Cambridge, and perhaps humanity, given his knowledge and intimate professional involvement in American preparations for nuclear war. The letter lays bare Conant's fear and anxiety about the future, a set of emotions that are not surprising, given his leadership as a chemist on committees that oversaw the Manhattan Project, which made possible the development of American capacity for nuclear war in the years leading up to this moment.

Also significant in the story we tell is the letterhead Conant writes on, and specifically the geographical coordinates mapped therein: official university letterhead from his office and official residence, addressed as only "17 Quincy Street, Cambridge, Massachusetts." The proximity of Conant's location to Whittlesey and Kemp's queer lives at The Loft at 20A Prescott Street is striking. The map in figure 6.3 depicts the proximity of these locations within the small city of Cambridge. One could walk the 200 meters from 17 Quincy to 20A Prescott in a matter of three minutes, with the Harvard Faculty Club, a social hub, situated directly between the two homes. We have, in fact, walked this distance in about three minutes, veering around the Faculty Club, through the lovely gardens that surround it, a convening point then and now that sits amid university-owned apartment and commercial buildings that densely line both streets. Conant's geographical proximity would make it nearly (seemingly?) impossible for

Figure 6.2
Photograph of President James B. Conant. Courtesy of Harvard University Archives, Harvard University Library.

him, and others, to *not* know something of the intimate relations unfolding nearby at The Loft.

In this chapter, we unpack compelling evidence of President Conant's centrality in terminating geography at Harvard, primarily through his retention, as college president, of the right to confer or deny tenure and promotion of members of the faculty. We argue that whereas Bowman chose to "remain silent" and treat politics surrounding geography delicately, Conant did not hesitate to wield the significant resources, power, and influence attached to the Office of the President to achieve whatever he believed to be morally correct—even when this placed him at odds with powerful entities in his midst, such as the Board of Overseers and Isaiah Bowman himself. As we will show, this included Conant's ability to decommission a field and, in so doing, to commission and influence outsiders, such as other powerful geographers like Bowman, to assess the legitimacy of geography at Harvard.

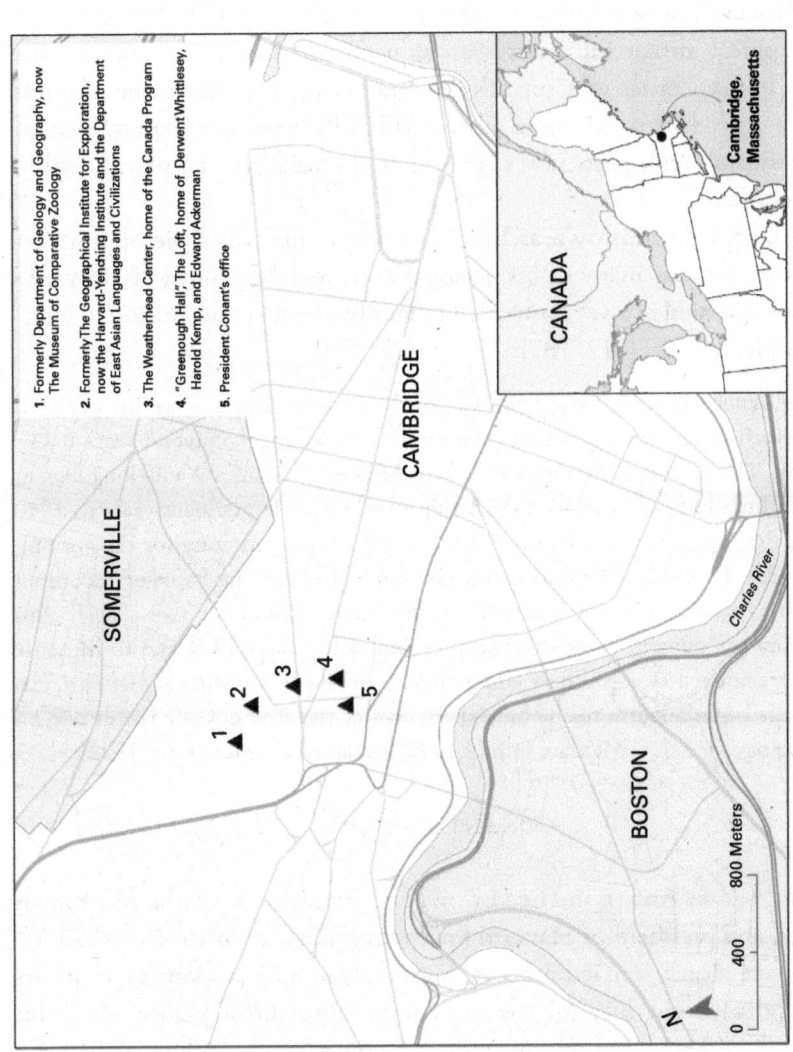

1. Formerly Department of Geology and Geography, now The Museum of Comparative Zoology

2. Formerly The Geographical Institute for Exploration, now the Harvard-Yenching Institute and the Department of East Asian Languages and Civilizations

3. The Weatherhead Center, home of the Canada Program

4. "Greenough Hall," The Loft, home of Derwent Whittlesey, Harold Kemp, and Edward Ackerman

5. President Conant's office

SOMERVILLE

CAMBRIDGE

BOSTON

Charles River

CANADA

Cambridge, Massachusetts

N

0   400   800 Meters

Figure 6.3
Map of significant locations in Cambridge, Massachusetts. Map by Trina King.

In this chapter, we reveal a series of contradictions that belie James Conant's positions. Conant's professional views are well documented in academic and popular sources (e.g., Hershberg 1993; Conant 2017; Hershberg 2019). Therefore, this frees us in the pages that follow to explore how Conant's dogmatic views informed his position and actions to end geographical instruction and research on campus. We also examine some of the idiosyncrasies of Conant's life and personality that make him not merely a product of his time (a time rife with homophobia and fear of communism) but a particular character in the roles that he played within these contexts.

With regard to his own archives, Conant seems rather the opposite of Bowman, keeping many of his personal notes and diaries out of Harvard's archives and hidden away, entrusted to family members, according to biographer Hershberg (2017, 187):

> One crucial item missing from Conant's Harvard papers were his personal diaries (in some cases glorified appointment books) for the pivotal years 1941–1950. They certainly *should* have existed. After all, Conant was a lifelong diarist, with little red books of jottings dating from his undergraduate years to 1940 and then resuming, in identical form, in 1951 and continuing for the ensuing decades. In the 1940s Conant did not suddenly drop the habit of recording events, as shown by the extensive handwritten accounts of two crucial trips during the missing years—to wartime England in early 1941 and to Moscow in December 1945—both of which survive in his personal files at Harvard. His detailed, scribbled notes of his experiences at the first nuclear bomb test, in Alamogordo, New Mexico, in July 1945, remained classified until 1982.

CONANT'S ORIGINS

Conant was an American chemist, major administrator of the Manhattan Project, and president of Harvard for twenty years, from 1933 to 1953.

Conant clearly carried a heavy burden, assuming responsibility as his civic and patriotic duty for rooting communists out of higher education by pursuing them on campuses across the United States. When he died in a nursing home in Hanover, New Hampshire, in February of 1978, the *New York Times* published the following headline on its front page: "James B. Conant Is Dead at 84; Harvard President for 20 Years" (Associated Press 1978). The article identifies Conant as scientist, educator, and diplomat.

Conant was born in Dorchester, Massachusetts, in 1893 and raised in a lower-middle-class home in Boston. This upraising set him up as an antagonist to the city's ruling class, a precursor for his clashes with aristocratic decision-makers who influenced the Board of Overseers that governed Harvard. Conant himself entered Harvard as an undergraduate student at the age of seventeen and graduated *Phi Beta Kappa* three years later, only to return a few years later, in 1916, to undertake doctoral work in chemistry. When he was confirmed as Harvard's twenty-third president, Conant was forty years old.

According to the obituary published in *The New York Times* in 1978, Conant told the following story at a dinner party at the Harvard Club of New York shortly after his own presidential inauguration at the college in 1933: "The situation in which I find myself recalls an experience of Sir William Osier, the physician, while touring in Canada. It was spring. The roads were very muddy. Sir William came to a signpost which read 'Choose your rut now; you will be in it for 35 miles'" (Associated Press 1978).

Although the letter opened by Harvard President Drew Faust in 2000 suggested uncertainty, not much else about Conant's time in office expressed incertitude. The lean, tall figure was known for his efficiency and decisiveness. His views on education were radical, if not popular, and indeed he endeavored to change public education in the United States well beyond Harvard's campus, targeting what he understood as compounding social issues in American cities.

The Associated Press characterized Conant as impervious to pressure or social criticism: "As an educator Dr. Conant was indifferent to the pressures of politics and consensus and, although he did not seek out controversy, he was unintimidated by it. He would cling obstinately to his opinions if he thought them right. On the wall of his office at Harvard he kept a framed cartoon with the caption: 'Behold the turtle, he makes progress only when his neck is out'" (Associated Press, 1978).

Joining the university in the thick of the Great Depression, Conant believed his educational mission was to promote perceived American values, challenge the influence of Boston's entrenched aristocracy at Harvard, and use social sciences to combat the spread of communism. Due to his belief that geography was not a rigorous scientific field, among other things, Conant opposed the expansion of the discipline at the university and, ultimately, its existence at Harvard at all. The presence of explorer

Alexander Hamilton Rice, who he saw as buying his professorship with the funds bequeathed by his wife to build the Institute he directed, reinforced this antagonism, and the presence of queer geographers on faculty likely cemented the field's demise on campus.

Conant was—first—a Harvard professor of organic chemistry. He held narrow views on science. While president of Harvard during World War II, Conant served as scientific advisor, contributor to creation of the atomic bomb, and member of the committee that selected Hiroshima as its first target in 1945 (Associated Press 1978).

After completing his presidency, Conant embarked on a new career as diplomat, serving as US ambassador to Germany.

THE SECRET COURT, "UNAMERICAN ACTIVITIES" AND THE "TEMPERAMENTALLY UNSUITABLE"

While Conant was not merely a product of his times, the times in which he lived were most certainly significant to his role and social location.

In 1953, two years—almost to the day—after Conant penned his speculative note to the future president of Harvard, notorious anticommunist US Senator Joseph McCarthy signed his own letter addressed to another president: that of the United States. This typed letter from McCarthy, in fact, references James B. Conant. McCarthy wrote the letter as a courtesy to the president, to explain the four premises on which he would oppose the confirmation of James B. Conant as high commissioner in Germany when it came to the Senate floor a few days hence. The four principled grounds on which McCarthy would proceed in opposing Conant's confirmation included Conant's advocacy of the destruction of industry in West Germany, his advocacy of the redistribution of wealth, his opposition to parochial schools (which would "furnish ammunition for the Communist propaganda guns" in Germany), and Conant's denial of the presence of communists at Harvard. On this last point, McCarthy names three Harvard professors: Harlow Shapley, Kirtley Mather, and the late F. O. Matthiessen. McCarthy notes that everyone believes these three to be or have been communists, with the exception of Conant himself.

McCarthy continues:

Let me make it clear that I do not accuse Mr. Conant of being either Communist or pro-Communist. However, I strongly feel that his innocent statements about

Communist activities in education and about the presence of communism in his own faculty indicate a woeful lack of knowledge of the vicious and intricate Communist conspiracy. Certainly it doesn't show any qualifications for the task of safeguarding the American embassy at Bonn against Communist penetration, nor with the ask of meeting the Communist threat in West Germany.[2]

McCarthy then notes that Conant seems "a fine gentleman" and one intelligent enough to have stated at a convocation address at the University of New York in 1947 "that he was greatly concerned by the fact that many temperamentally unsuitable persons were making their way into universities."[3]

While McCarthy did not make these views public, only voting rather than speaking against Conant in the end, they are nonetheless ironic, given Conant's own concern and role in removing communists and the "temperamentally unsuitable" from campuses across the United States.

The hysterical fear of communism that had taken hold among American politicians and the public was layered upon and intersected with another social hysteria: the fear of queer people and their sexuality. Same-sex relationships were criminalized, with queer people ostracized informally and formally, sometimes through explicit policies to remove them from society. This was certainly the case within Harvard's gates where, in the 1920s, a "Secret Court" of administrators was formed to remove queer people from campus, resulting in fourteen students being expelled and two subsequent suicides among them (Wright 2006; Cho 2021). Celebrated Dean Greenough, after whom the building containing The Loft was eventually named, convened the Secret Court. In 2020, a group of student activists convened to pressure Harvard to award posthumous degrees to those students the earlier administrations had expelled, as one of several steps to reconcile with this past (Cho 2021).

Homophobia operated virulently in all mainstream institutions at the time. The ranks of the US federal government, for example, were also being purged of homosexuality, a destructive campaign that would continue through much of the rest of the twentieth century. The Lavender Scare was a moral panic about queer people in the US government, which led to their mass dismissal from government service during the 1950s. According to David Johnson (2006), the scare reflected how the federal bureaucracy came to be thought of as a haven for socialists, misfits, and perverts. Yet, as Johnson writes, this history was also the story of the effect of the purges on local queer populations—and the chilling effects of government witch hunts thereupon. Johnson argues, however, that it was precisely this persecution of queer people that led to the rise of the first organized opposition to the federal government's antigay policies. Though the US federal government created such policies to contain a perceived, growing homosexual menace, the Lavender Scare inspired not only the founding of a gay movement but also the later radicalization of that movement in 1960s Washington, DC. This came, as Johnson identified, before the pivotal 1969 Stonewall riots and led to the development of many of the future organizational tactics and legal victories underpinning the gay rights movement. These events worked in parallel with and functioned to support more specific instances of homophobic persecution at this time, such as Florida's Johns Committee's witch hunts against students and scholars at its educational institutions (Graves 2009; Braukman 2012).

Related purges placed communist-leaning leftists, queer people, and general campaigners for social justice into the same category, at a time of intense societal pressure to conform in the mid-twentieth century. This was the time in which the queer geographers lived, hidden, at the center of campus, and professors were considered particularly "temperamentally unsuitable," in the words of Conant himself in a speech he gave where he referenced the challenges associated with the personalities and politics of college professors.

Social norms and heteronormative, classed, and racialized identities were oppressively policed at the time in which Conant, Bowman, Whittlesey, and Kemp lived. Within these societal contexts, Harvard functioned as an elite institution invested in maintaining the broader societal status quo, regardless of the intellectual work, philosophical beliefs, or sexual activities of its students and faculty. These cultural contexts and social forces intersected to make Harvard's campus ever smaller for queer people like Whittlesey and Kemp, whose love placed them outside the realm of acceptable behavior and identity, placing them at very likely risk of losing their livelihood were they to be outed.

CONANT'S PHILOSOPHY OF EDUCATION, AND THE "PURITY" OF SOCIAL SCIENCES UNENCUMBERED BY RADICAL IDEOLOGY

Conant was known as a pragmatic man who believed in the importance of science to world events (Biddle 2011). Histories of the Manhattan Project place him at the center of the project, as leader and influential member of the Military Policy Committee that convinced President Roosevelt to invest in the development of atomic energy as weapons of mass destruction (Associated Press 1978).

Once he returned from his diplomatic post in Germany in 1957, however, Conant devoted his professional life to what is often referred to as "the democratization of education," first by leading research funded by the Carnegie Foundation to improve American public schools, then understood to be in direct competition with the Soviet Union.

For someone who believed that academic freedom means that scholarship should happen unencumbered by state influences, one could certainly also observe that Conant's work was heavily influenced by the anticommunist politics of the time. In a conservative climate rife with virulent American nationalism, Conant believed that social science should be dedicated to the good of American values.

During his time as president, Conant's views are readily on display in two books that he authored on the topic in 1948 and 1953. In the first, *Education in a Divided World,* Conant (1948) argues that the United States and the "free world" are in an ideological conflict with the Soviet Union, with an attendant domestic battleground being the minds of the masses, pitting "Soviet philosophy" against American values such as freedom. For Conant and other leaders, the calculus of the war depended not solely on weaponry but on "our will to develop the physical and spiritual strength of our people as a democracy" (1948, 2). This meant propagating a common body of social values and mores among "the American people": individualism, private property, profit, meritocracy, freedom of opinion, and equality of opportunity (1948, 6). Conant understood public education as the locus for transmission of these ideas:

> The impact of European radical doctrines . . . based on the notion of a class struggle . . . have to a considerable degree diverted the attention of forward-looking men and women from the social goals implicit in our native American traditions. As a consequence, we have thought too little of our system of public schools . . . as an instrument of national policy. (5)

Conant believed that children required indoctrination in pursuit of national solidarity, making public education Conant's bulwark against communism. Although Americans can easily defeat Soviet philosophy in a "free market of ideas," they must establish it first (22). Once built, the market will create greater social mobility for all American youth, thereby leading to what he calls "equality of opportunity" (7). Conant believed that the lack of equality of opportunity, not social injustice, led to communism. Conant posits those who disagree with this liberal utopianism as "taking their orders from Moscow" or "being sympathetic to a stratified social system" (i.e., radicals or reactionaries) (10). With the free market as a source of political legitimacy, these are the premises from which he argues how to structure education to promote "American democracy."

Conant found that not all values were suitable for children, but rather only those which upheld strict individualism based on "Hebraic-Christian" ethics fit for a free society (103). Marxism, on the other hand, is filthy, corrupt, and immature; Conant likens its spread and invisibility to pornography on an "academic black market" (173).[4] This virulent image was intentional and also common at the time, with the notion of communism spreading like a cancer. For Conant, then the academic must swear himself to a fervent defense of individualism against the collectivist disease (176): "our philosophy is superior to all alien importations" (173), ideological superiority can only be maintained through purity—meaning that no circulation of radical ideas—or people—should occur (180).

Conant argues that the mass mobilization of free, public education is necessary to wage this ideological war (41; 48), with American values taught in "general education" as the core curriculum for most students—including half of four-year university students, who will be moved to two-year colleges (154–156). While Conant suggests that general education should include social science, he also believed it ought to be confined to the distillation of American values: "If coupled with practical experience and infused with a zeal to move American society along its historic road, [the social sciences] may be particularly effective" (36). What we would today identify as critical social science should be taught only at very high levels of education, which would effectively remove it for the vast majority of Americans (199). Conant justifies this by stating that these people are those from whom "society has much to fear," because "from such people come the leaders of anti-democratic movements." Thus, Conant develops a general education to produce individuals with American values, tying social sciences to this cause.

Geography makes almost no appearance in the pages of Conant's book, not even in the list of subjects of the social sciences or sciences. Geographical knowledge was considered a necessary part of general knowledge but only insofar as it can produce socially effective Americans. Conant intends to root out scholars not committed to American values by forcing them to "declare their own basic social philosophy," adding that the issue primarily applies to the social, not physical sciences (175). The creation of a Department of Geography in the late1940s would have clashed with his stated purposes of consolidating university education and reorienting the pedagogy of the social sciences. It would also conflict with the likely presence of radical ideas held by Harvard personnel apt to teach subjects that did not align with Conant's vision of the liberal social scientist. The purpose of this scientist was not to conduct free and open enquiry but to shape American values and solve American problems using scientific tools. Conant argued repeatedly that the ideal disciplines to achieve this end were education, sociology, anthropology, and social psychology, which could understand and alter human behavior through education. Perhaps geography simply had no place in this reimagined social science and no place within Conant's ideas of science. If he saw no place for geography in this project, then he was consistent in not supporting its expansion in the university curriculum he sought to consolidate.

In his subsequent book, *Education and Liberty* (1953), Conant proposes a comparative analysis of education systems of the United States, the United Kingdom, Australia, and New Zealand, selected by virtue of their common Anglo-Saxon heritage. The short analysis relies on basic descriptive statistics and interviews with educators, giving way in text to Conant's opinions and hopes for the future of American education. Conant views the American educational system as successful, arguing that its universal, public, and decentralized characteristics promote core American values, which he aligns with "Jeffersonian democracy" (e.g., protection of property, profit motive, right to assembly, universal suffrage). Higher-level education is not simply for vocational training but for turning all Americans into citizens; to produce these citizens, it is necessary to teach them "general knowledge." Universal education also creates "equality of opportunity" and a diffusion of cultures, religions, and 'races,' which Conant states will remove political divisions. It is on these bases that he argues for the expansion of higher-level education for all Americans, including increased teaching of the sciences and mathematics, additional support to the "gifted," enrolling more students, and providing more standardized curricula in high schools. For universities, it requires the contraction of four-year programs in favor of two-year programs that would teach general knowledge (akin to the "liberal arts") in contrast with more detailed, vocational knowledge in four-year programs.

Conant's attacks on communism are scathing, positioning American schools as battlegrounds for ideological war:

> It may well be that the ideological struggle with Communism in the next fifty years will be won on the playing fields of public high schools of the United States. That this may be so is the fervent hope of all of us who are working to support and improve these characteristic American institutions. (62)

He places securalism at war with religion (79). Religious groups implicitly support private schools and, therefore, pose a political challenge to Conant's push for public-based education. He attacks religious schooling on its economic limitations and claims to primacy in teaching morality and spirituality. Conant argues that, by emphasizing private, denominational schooling, many talented students[5] will be left behind and that public school support does not exclude private schooling. He believes that morality and spirituality, through Jeffersonian ideals, can and should be taught in public schools; in particular, being public does not prevent schools from teaching spiritual ideas in a nondenominational way.

State control or influence is another core challenge to American education. Conant particularly denounces Franklin D. Roosevelt's administration, especially through the National Youth Administration (NYA). Given that he sees decentralization as a mode of experimentation through markets, heightened central authority is both inefficient and authoritarian. (Conant even likens centralized education to the implementation of communism.)

Assuming that Conant's values in 1953 reflected those he held as president of Harvard during the twenty years preceding publication, there are links between his proposals and the end of geography. He generally favored reduction in four-year college programming, especially for programs that used state support.[6] He opposed programs (and, thereby, likely people) that would not teach what he saw as American values. He opposed higher-level programs that he saw as too easy (i.e., not scientific enough). He opposed the proliferation of publicly funded state universities. It is possible he associated these characteristics with geography in the late 1940s.[7] The creation of a Department of Geography would expand four-year programs, bring in academics of different values, and not be scientific, in Conant's perspective, due to the focus on human geography. State universities were chief sites of the creation and expansion of geography departments, which could have led Conant to conclude that it had sufficient higher-level coverage, or to associate it with state influence. Conant, moreover, supports teaching geography as part of American education in the book.

This understanding of Conant's views on education, while recorded in writing after the closure of the Department of Geography, sheds light on and sets the stage for deeper understanding of his views on geography not being a robust science and on his questioning of the scientific legitimacy of geography.

CONANT'S OPPOSITION TO GEOGRAPHY

The Geography Program is something that Conant inherited, seemingly reluctantly, when he became president. It had been the Lowell administration that showed the vision and took action to recruit Blanchard and Whittlesey, if not Bowman (although they tried). In 1927, Conant's predecessor, President Lowell, received a memo from then dean of the Faculty of Arts and Science (FAS), about a position that paved the way to build a Department of Geography on campus, beginning with the decision to hire two men suited to build this new program, thus tying Whittlesey's fate to that of the program from the outset. The dean stated: "The Senior members of the Division of Geology are heartily in favor of asking Blanchard to come for a series of years—two or three at least—for half of each year. This plan presupposes inviting a young man as Instructor, or Assistant Professor at the most, who may come under Blanchard's influence and, we hope, develop so as to carry on the work when Blanchard feels that he must drop out." The Dean then explained that he would be lunching with Blanchard at "The Colonial Club" that afternoon and wished to offer him the position and proposed compensation.[8]

Part of President Conant's belief in the purity of social sciences was that certain among them lacked purpose and scientific rigor: namely, geography. Harvard University Archives and Bowman's archives at Johns Hopkins pulsate with Conant's dismissive views of a discipline he viewed through a masculinist lens as insufficiently scientific. There is evidence of geographers on and off campus attempting to persuade Conant of geography's merits, most prominently and perhaps intimately through direct correspondence with Isaiah Bowman, whom he addressed as "Jim." In 1942, with the war under way, Bowman wrote to Conant to acknowledge receipt of an article that Conant had sent him titled "University Training and War Service in Great Britain."[9] In the same letter, Bowman then reported to Conant about a special committee that was forming in New York, comprising scholars from a number of campuses to advance these discussions, and praised the scholarship of the individual with whom he was collaborating to organize the committee.[10] It was also evident, as a subsequent example, in Bowman's extensive notes on the fateful events on campus that ended geography in October of 1948, that Bowman and others on the board well understood Conant's refusal to conceive of geography as a legitimate science. In his recollections recorded October 13, 1948, two days after the fateful board meeting, Bowman described in detail his conversation in a corner with Conant and a Columbia physicist explaining the rotational deflection of Long Island's valleys: "My purpose was to show Conant that geography has as good examples as chemistry—and as much claim to consideration as a science."[11]

But repeated attempts to prove geography's scientific merit over the years had already failed. When he died, Conant was memorialized frequently as an uncompromising person. He did not hesitate to wield his power and influence to achieve ends he believed to be principled and morally correct. We found this to be the case in the way that he involved Bowman in decision-making about Ed Ackerman's tenure file and about the Geography Program. As noted in chapter 5, the two became close friends, with Bowman staying at Conant's home, and Conant invited to Bowman's home six months after the conferral of his honorary degree at commencement in Cambridge in 1936:

> I hear that preparations are afoot to have you attend a luncheon and dinner in Baltimore. . . . I write to ask if you and Mrs. Conant (if as I hope she is to accompany you) will not be good enough to spend the night at our house. We can make you comfortable and it would be great fun to have you. . . . My wife will write to Mrs. Conant just as soon as we hear if it is possible for you to come.[12]

Conant leveraged his friendship with Bowman, knowledgeably stroking Bowman's ego, desire for stature, and influence over the years, offering

Bowman not only his friendship but an honorary degree in 1936 and eventually a seat on Harvard's Board of Overseers in 1948. In the intervening years, they cemented their friendship, and their correspondence became more familiar. When Bowman's office mails a form letter to Conant's in a bid to request speculative postwar funding forecasts for "research expenditures," Conant feels comfortable enough in his rapport with Bowman to declare, "I frankly think the inquiry is unadulterated nonsense," akin to "the famous remark about the Scotchman whose capacity for drinking whisky was estimated as 'any given amount'!"[13] Conant ends his letter: "I realize that all of the above is none of my business and I am presuming on our friendship to give this off-the-record statement of my reactions to your official letter."[14] The two then go on to exchange jocular letters that blend the temporary illnesses that confine them presently to home (laryngitis for Bowman and a bad back for Conant), where they continue to type letters to one another about university business such as soliciting questionnaires and finding scientists to serve on university committees.[15]

There would have been little doubt, then, when Conant wrote to Bowman in September of 1947, asking him to serve Harvard on the external committee that would review Ackerman's tenure and promotion application, that Bowman would agree to this request. Conant, by this time, likely knew exactly how to appeal to Bowman's sense of self-importance, even when he also likely knew that the two would disagree on the fate of geography. The letter soliciting his involvement ends with the following: "I apologize for all this trouble. I can only plead the excuse that Harvard needs your assistance very badly, and I am venturing to presume on your good nature."[16] Conant used the phrase "presuming" often, significantly, likely knowing that the friendship the two had cultivated would bring Bowman along in the not-to-distant future.

UNFINISHED BUSINESS: FIVE INSTITUTIONAL REVIEWS
OF THE CLOSURE AFTER 1948

The immediate consequences of implementing President Conant's policy to let geography die were to remove all nontenured instructors and end geography as concentration. The final condition leading to geography's decline was its gradual loss of resources. The administration removed geography as a field of concentration by March of 1948. Whittlesey summarized the situation in April of 1948, in the immediate aftermath of the decision, in a letter to colleague Willard Miller:

> Its immediate effect is to eliminate Geography 1 and two other courses given by Richard Logan. Ackerman and Ullman are continued for one more year. . . . The net effect is to eliminate geography as a field of undergraduate concentration after the present junior class is finished. . . . However, the really great resources of Harvard, including its library and the building constructed for geography, will largely be useless.[17]

The decision to not promote three faculty members, the elimination of the concentration, and the reduction in courses all contributed rapidly to the field's elimination by the end of April of 1949. Although Ackerman and Ullman had their jobs "continued" for one more year, Ullman was unable to teach in geography in 1949 due to his commitments in Regional Planning, and Ackerman resigned in August of 1948.[18]

Likely overwhelmed with stress, Whittlesey listed geography courses lost due to Conant's policy in another letter to colleague Robert Johnson in November of 1948.[19] The convergent and spiraling nature of these events in quick succession caused Whittlesey to become overloaded with work, and he, therefore, had to decline teaching geography's introductory courses (101a and 101b) in 1949.[20]

Physical geography, already in decline, effectively ceased upon Professor Kirk Bryan's death in 1950 (Harvard University 1950). While Harvard nominally could have transferred the Institute's infrastructure and capital to the Geography Program after its closure in 1951, the administration instead removed the Institute, sold its machinery, and gave the building to the Department of Mathematics by 1953 (*Harvard Alumni Bulletin*, October 13, 1951; Harvard University 1954). Combined with the refusal to reverse policy, specifically by allocating no new funding, personnel, or infrastructure, the Geography Program lost its ability to continue research and instruction by the end of 1956. New losses followed previously described cuts, which had begun in the 1940s. They included the removal of all nontenured professors and instructors by 1949, the end of geography as concentration, the removal of all tenured professors and permanent courses by 1956, and a formal end of the graduate program by 1960.

In 2025, as we write, geography itself remains a problem that hangs open at Harvard, characterized in the archives and in life, as in its untimely death on the campus, as "unfinished business." Like most geographers, we knew little of what happened after its closure. During the course of our research, we learned that various people endeavored to revive geography over the years, including, most recently, Professor Peter Bol, founder of the Center for Geographic Analysis and occupant as faculty member of the building formerly known as the Geography Building.

Multiple committees reviewed the situation of geography at Harvard beyond 1948, including a secret 1952 report by London-based geographer Dudley Stamp (1952), commissioned by none other than President Conant and ultimately calling for its revival (a call that was suppressed). We explore these five reviews and discuss Rita Morris's (1962) account of this history, which connects the death of geography at Harvard to wider trends and the fate of geography as a discipline, likely promulgating geographers' long-lasting fears that these two histories were entwined all along.

President Conant's policy to let geography die ultimately led to its complete demise by 1959. This was possible only due to the continued, active implementation by the Conant and Pusey administrations of the "let geography die" policy. This continual decline involved four necessary, joint conditions:

1. The Administration's refusal to consider or accept review of its policy with respect to geography;
2. The College's refusal to allocate new resources to geography;
3. Geography's total loss and reallocation of its existing resources; and
4. Derwent Whittlesey's death in 1956.

These conditions slowly abolished the program's resources until it no longer had personnel, infrastructure, or funding to continue research or instruction.

The Conant and Pusey administrations refused to consider or accept review of the policy to let geography die.

Nonetheless, we found evidence of at least five institutional reviews of the status of geography at Harvard between 1947 and 1956, and we consider these reviews as afterlives of geography on campus.

First, as noted earlier, the external ad hoc committee on Ackerman's promotion, chaired by Bowman, overstepped its mandate to review Ackerman's case and recommended the creation of the Department of Geography in December of 1947. Second, the Board of Overseers conducted a unanimously approved review in summer of 1948 that argued that President Conant lacked the authority to implement his policy and that the matter must be forwarded to the FAS for further review (Board of Overseers 1948).[21] Third, having received this directive from the board, FAS created a subcommittee on the issue that met weekly for a year, concluding in 1950 that Conant's policy should be reversed and the Department of Geography created (*The Harvard Crimson* May 11, 1949).[22]

Fourth, President Conant himself commissioned London School of Economics geography professor Dudley Stamp to review the situation of geography two years later. In his report, completed in 1952, Stamp concluded that Conant's decision should be overturned and geography revitalized using the Institute's infrastructure (Stamp 1952). This is, by far, the most intriguing review, not only for its initiation by Conant himself but for the secrets, findings, leaks, and controversies associated with it. Stamp was a well-known geographer based at the London School of Economics and was in touch with both Conant and Whittlesey, separately, about this review.[23]

In his report, Stamp argued that four things were needed to support geography at Harvard: suitable accommodation, specialized equipment, a library, and additional staff. Stamp observed that only additional staff were missing from these four requirements. Referring to the Institute he wrote, "I am familiar with all the Universities [sic] of Canada and a very large number in the United States and Europe, but I cannot recall a building more suitably designed for the purpose of a Department of Geography" (Stamp 1952). He added that there was an urgent need for geographical knowledge, writing, "so many of those with whom I discussed the problem [of geography] expressed the need of their departments for aid from the geographical side" (Stamp 1952). In a confidential appendix, however, Stamp leveled a secret critique of Whittlesey despite their common interest in reviving geography:

> For many years I have enjoyed the personal friendship of your present Professor, Derwent Whittlesey. He might well say that the program I have outlined differs but in detail from what he planned and would have carried out. His shy retiring disposition and dislike of administration have prevented his qualities from being appreciated. (Stamp 1952)

Finally, a committee on the "behavioral sciences" at Harvard reviewed geography's position in June of 1954. The committee classified geography under "ecological and demographic subjects" and made comments on its situation at the university despite it reportedly falling outside of the report's scope. Here, the committee reflected on interviews with anonymous participants, stating:

> Our inventory of work in geography, demography, and related subjects shows widely scattered centers of activity. Certainly nothing approaching a major, integrated program in these fields now exists at Harvard. This situation was regretted by several of our informants. The "de-emphasis" on geography is deplored by some; others feel that locational and regional studies in economics have an importance far transcending their representation at Harvard or in the country at large . . . The desirability of an integrated program which would bring together the now-scattered foci of interest was stressed by these men. (Harvard University 1954, 72–73)

While analyzing previous reports calling for reviving geography at the university, the committee ultimately concluded that "the gap in geography at Harvard . . . is one of the University's serious unresolved problems" (Harvard University 1954, 73).

All five reviews recommended reversal of Conant's decision. Despite these repeated reviews and calls for geography's revival, the college refused to change course. As he approached the end of his term, Conant chose to defer any decision until the election of a new president.[24] Ultimately, the new administration of President Pusey rejected the findings of these reports unless geography received a new, large, and private endowment.[25]

Another reality guiding geography's decline was the administration's refusal to allocate any new resources. As noted, President Conant's original declaration explicitly referenced limits to supporting geography due to "demands on funds."[26] We also showed that this policy implicitly continued a trend of refusing to find funds for geography since the 1930s. This rationale and practice, moreover, persisted until the program's closure in 1958, despite consistent findings and calls to reverse the policy.

President Conant maintained his decision to allocate no new resources, even when the Institute desperately needed them and when its newly available resources placed geography in an opportune position for rebirth. The reason for the Institute's closure in October of 1951 was its lack of funds. Hamilton Rice personally sought out Conant to transfer the Institute to Harvard's direct control and continue its funding; however, the Administration claimed that it would be unable to find funds for that purpose.[27]

It also publicly declared that geography could not be revived without at least "funds to underwrite the minimum of two additional professorships."[28] Conant estimated their joint cost at between two to three million USD with an annual operating budget of nearly 400,000 USD—each far more than the annual or total operational costs of the Institute during its lifetime.[29] This was despite the fact, for example, that Stamp had independently determined that a budget of 93,000 USD per year would be sufficient for the new Department of Geography.[30]

In 1952, when Stamp initiates his research at Harvard, by request of Conant, Whittlesey writes to him during his stay at the Hotel Statler in Washington, DC, to follow up on a phone call they had by arranging his visit, along with Mrs. Stamp, to Cambridge in September, including accommodations. He then thanks him for his work to revisit and potentially sustain and nourish what remains of geography including, still, the Geography Building:

> May I thank you again for your willingness to undertake this task. I hardly need repeat what I said on the telephone, namely that we have no funds at present for expanding our work in Geography, but before we assign a building formerly used as the Geography Institute to other purposes, I think we should make an effort to raise a modest endowment for Geography. Just what that modest endowment would be and what program it would support is the problem on which I am asking your advice. It is hardly possible to go to a foundation or a donor with only a vague notion that something should be done about Geography. Therefore, you will see that your report will be an important step in clarifying an issue. With deep appreciation of your interest and your willingness to assist, I am.[31]

News of Stamp's review was leaked to the press, prompting Stamp to mail to Conant an apologetic letter and the clippings, shown in figure 6.4. In his reply, Conant was departing the presidency to take up his new position in Germany. He assured Stamp: "The fate of geography will lie in my successor's hands, and he will pick up where I have left off."[32]

THE FINAL REPORT

One year after Conant's time in office ended, yet another report was issued on the status of geography. This fifth and final report notes six doctoral students in geography between 1948 and 1953, probably holdovers—students

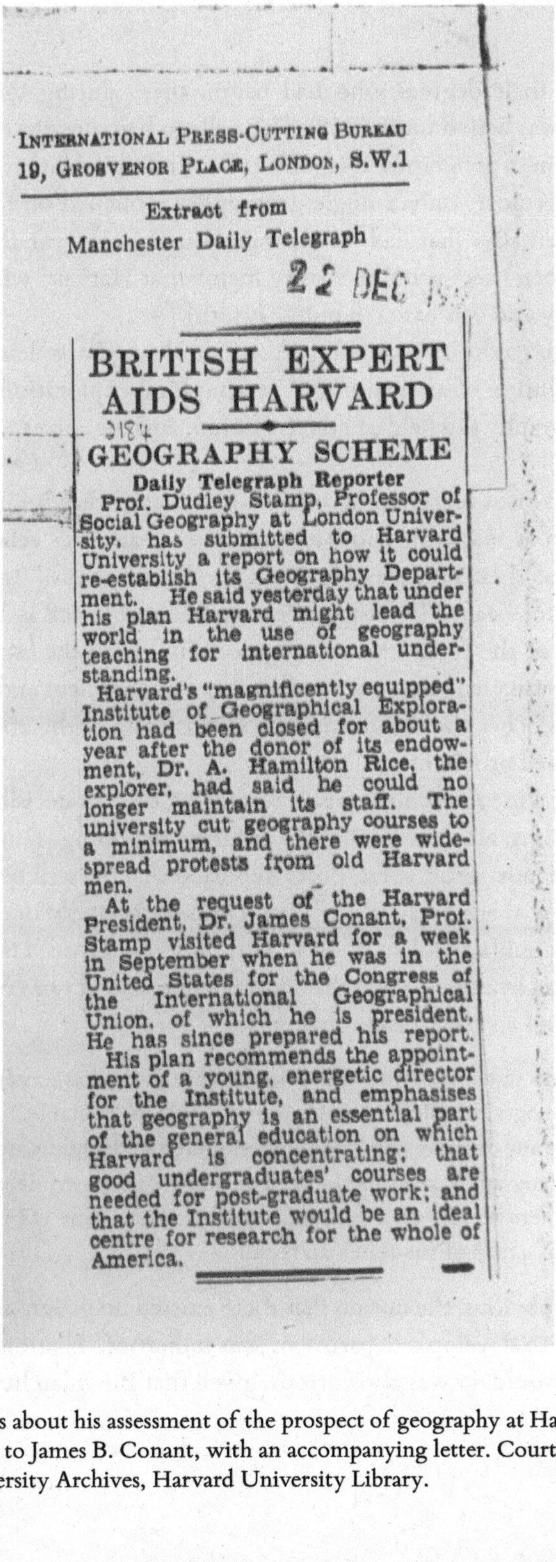

Extract from
Manchester Daily Telegraph

2 2 DEC 19

BRITISH EXPERT
AIDS HARVARD

218₄

GEOGRAPHY SCHEME

Daily Telegraph Reporter

Prof. Dudley Stamp, Professor of Social Geography at London University, has submitted to Harvard University a report on how it could re-establish its Geography Department. He said yesterday that under his plan Harvard might lead the world in the use of geography teaching for international understanding.

Harvard's "magnificently equipped" Institute of Geographical Exploration had been closed for about a year after the donor of its endowment, Dr. A. Hamilton Rice, the explorer, had said he could no longer maintain its staff. The university cut geography courses to a minimum, and there were widespread protests from old Harvard men.

At the request of the Harvard President, Dr. James Conant, Prof. Stamp visited Harvard for a week in September when he was in the United States for the Congress of the International Geographical Union, of which he is president. He has since prepared his report.

His plan recommends the appointment of a young, energetic director for the Institute, and emphasises that geography is an essential part of the general education on which Harvard is concentrating; that good undergraduates' courses are needed for post-graduate work; and that the Institute would be an ideal centre for research for the whole of America.

Figure 6.4
News clippings about his assessment of the prospect of geography at Harvard sent by Dudley Stamp to James B. Conant, with an accompanying letter. Courtesy of Harvard University Archives, Harvard University Library.

completing their degrees who had begun them during the years when geography was active until 1958. The college had already eliminated the concentration in geography as an option for undergraduate students by this time. Furthermore, only a single geographer remained on faculty (a situation, incidentally, that had not changed during the year that we visited Harvard, when there was one faculty member at Harvard who held a PhD in geography and was based in public health).

According to the 1954 report: "A recent 'deemphasis' has been marked by the dissolution of an institute of geographical exploration, the elimination of geography as a field of concentration, and the corresponding reduction of staff."

This status was regretted by many colleagues with whom we discussed the situation at Harvard, with contemporary academics echoing how the frustration of their predecessors about the "deemphasis." In 1950, a special committee headed by history Professor Donald McKay made a report to the dean of the FAS, wherein they recommended the establishment of additional permanent chairs dependent on an endowment and the ability to find scholars. This, the committee stated, was "one of the university's serious unresolved problems."

President Pusey's administration retained Conant's decision to issue no new allocations, albeit with slightly different reasoning. During his initial refusal to support geography, Pusey reported that "it will be quite impossible to renew work on a scale of distinction without obtaining substantial funds"—one million dollars.[33] FAS Dean Bundy reinforced the administration's position by arguing that the department could not be created without new funds and an appropriate chair:

> There seems to be some question in his [Bundy's] mind as to whether a department of geography [sic] in the University is really desirable. . . . He also said that before any drive for funds could be mounted with enthusiasm, there should be some geographer available to head up such a proposed department. In his opinion, there was no such outstanding geographer in the USA today. He was thinking of a man of the Bowman type![34]

To state the obvious, the notion that there existed no geographer of the caliber needed to develop a department was ludicrous. That only "the Bowman type" would do was also curious, given that Bowman himself had first turned down the invitation to create the department and then failed to do so in his powerful position as a member of the Board of Overseers with direct access to Conant. Harvard continued to allocate no new resources until after Whittlesey, the last remaining professor in geography, retired and then died in November of 1956.

CONCLUSION: JAMES B. CONANT'S MANY CONTRADICTIONS

Some would argue that Conant abused his power in shutting down geography at Harvard. This position was contradictory, in its occupation by the man who fought to preserve academic freedom and intervention from the state, publicly in countless lectures and written statements. However, he also failed or refused to recognize that in the absence of state intervention, powerful institutions like Harvard and people leading them, like Conant himself, no doubt infringed on academic freedom through their actions, such as shutting down a field of research and instruction.

Another contradictory position Conant occupied was, having created the atomic bomb and leading the Manhattan Project, he then spent his adult life worrying over the potential destruction of society. Decades after Conant's death, his granddaughter, Jennet Conant, wrote a book about her grandfather—*Man of the Hour: Scientific Warrior*—noting that his efforts to regulate nuclear weapons had failed (Conant 2017). This fear was on display in the ominous letter he wrote to Harvard's future president, in which he speculated about the destruction of Harvard and Cambridge due to atomic war.

Equally contradictory was Conant's position on democratizing education to bring educational resources to the masses of American society, while working to narrow what could be included in educational content—especially in the social sciences—at what was arguably its most elite institution of higher education.

Also contradictory was Conant's view that geography did not serve American values as a weak social science, while virtually all geographers at Harvard were engaged in war work during his presidency. To be sure, Conant, Bowman, and Whittlesey were all putting university research and education in service of imperial war but with different interpretations of the place of geographic research and education in serving nationalist ends.

Despite these many contradictory views he held simultaneously, and which drove his professional life forward, President Conant was the right combination of dogmatic, powerful, and stubborn to have his views—contested as they were—last well beyond his own administration and for decades to come. Although challenged by five reports, James B. Conant's decision has lasted seventy-seven years at the time of publication of this book.

TALKING WITH GHOSTS

Simple? Why this is the old woe o' the world;
Tune, to whose rise and fall we live and die.
Rise through it, then! Rejoice that man is hurled
From change to change unceasingly,
His soul's wings never furled!

That's a new question; still remains the fact,
Nothing endures: the wind moans, saying so;
We moan in acquiescence: there's life's pact,
Perhaps probation, do *I* know?
God does: endure His act!

Only, for man, how bitter not to grave
On his soul's hands' palms one fair, good, wise thing
Just as he grasped it! For himself, death's wave;
While time first washes ah, the sting!
O'er all he'd sink to save.
—Robert Browning, lines 66–80

The gentle, perceptive counsel of this modest but wise man remains vividly
with many who shared his conversation and correspondence. Always a
gentleman, ever the scholar, his works toward international understanding
deserve the respect of his country, and grateful memory by his friends and
colleagues.
—Ackerman 1957, 445

These words, written lovingly, ended the obituary where Edward Ackerman remembers his cherished friend, colleague, mentor, and one-time
housemate, Derwent Whittlesey. Whittlesey fought strategically and for

years, as witnessed and documented in this book, to more deeply understand and restore to the public record and collective geographical memories
all that he sunk to save. Ackerman would have written these words in the
early stages of bereavement, for they were published only seven months
after Whittlesey's death, which Ackerman describes in his short essay as
"peculiarly untimely" (1957, 443).

Peculiar, indeed. For us, as for Ackerman, Kemp, and their colleagues,
there were likely also always two deaths entangled: that of geography at
Harvard and that of Whittlesey himself. We can only speculate, based on
the emotional distress detailed in letters by Whittlesey, Kemp, McSweeney, and others, that the stress and strain of the first death in 1948 might
have hastened the second, eight years later. The story, to us, is heart-
breaking, for the strain at work, and in his relationship, as detailed by
McSweeney, would have also broken Whittlesey's heart, emotionally and
physically.

In the decades since, American geographers have also been heartbroken
in their own ways, troubled, and most certainly haunted by these deaths,
by our collective failure to understand them and our collective inability to
confront them.

In this chapter, in service of the work of understanding, healing, and
contending, we continue our conversations *with* ghosts and *about* ghosts.
As theorists of haunting like Avery Gordon and Grace Cho write, there
is something unspeakable and, therefore, challenging about haunting to
social scientists who traffic in particular kinds of empirical evidence. In
contrast with more traditional social scientists, scholars of haunting cre-
atively explore possibilities for writing about the spectral, that which is not
seen or spoken, which is fleeting and in the shadows.

It is significant to us that our efforts to have conversations with the last
remaining living people that we knew of who knew Whittlesey and Kemp
in their time, for the most part, failed. Repeatedly, we reached out to our
elders, to senior scholars who were graduate students on campus at this
time, the last living students of Whittlesey's, only to find that they died
around the time of our outreach. In one case, Alison caught Whittlesey's
former student John Augelli on the phone, Kira seated beside her listening
in, both eager. We knew from a colleague that the retired professor was ailing at the time. Instead of speaking with us, John asked to have the call at
another time but died before that happened.

Funnily enough, in our lengthier telephone call generously granted by Olav Slaymaker, and in subsequent email exchanges after we were able to send him our work, Olav was dismayed by how little he knew of this whole history, even though he arrived at Harvard in 1961 to begin his master's degree in geography, only to find after traveling to Cambridge by boat from England that there was no longer a geography program to study in.

What we were left with, therefore, was more silence and our own frustration of living memories dying. Collectively, as geographers, we are left with our conversations with ghosts. Writing this book, for us, involved talking with ghosts through their archives and photographs, in buildings and memories, the spectral left behind but still very much alive in its own way.

We opened this book with one of queer American poet Walt Whitman's well-known poems, "Continuities." There, Whitman, too, suggests that nothing ever actually dies or disappears but lives on. Whitman's continuities speak to Avery Gordon's writing about haunting, to the notion that an oppression will linger and make itself known until properly contended with. Whitman's lines also articulate the life that can take root and rise from the ruins, come spring.

We suggest that neither Harvard, as an institution, nor geographers, as a vast personal and professional collective, have fully contended with this history, silenced all this time. This book is an opening, planting seeds for this reckoning. We offer these pages as an invitation to engage this history, too long suppressed.

As we research and write, we explore and, in some ways, live the connections and continuities between Derwent's personal and professional life and our own—not only in the work that we do as political geographers, but the subject positions and geographies that we inhabit. We have illuminated these seemingly chance encounters and continuities where possible. The relevance of our positionalities emerged poignantly for us at key moments during the research and writing of this history, during our own outings to one another, for example, and in demands by reviewers of an article about this history who questioned whether Whittlesey and Kemp were a couple and whether Ackerman—who later married a woman and had four children—was actually queer. Here, we continue our conversations with these reviewers and with Derwent's ghost as we map the haunting afterlives of geography at Harvard today, seventy years after his sudden death.

We are also compelled by Whitman's notion that the body, in death, gives way to spring when winter melts. Respectfully, and lovingly, we know that Derwent's body and body of work—and importantly, if intimately, his body of archives—has always had treasures to offer, sealed away for fifty years.

Whitman's continuities persist in a form of haunting at work in the lasting intergenerational effects of oppression that has not been spoken or dealt with, which, therefore, persists into the present. We find that this history haunts geography and geographers in the present in the discipline of geography in the United States, Harvard's campus and curriculum, and the queer geographers who arrived there, in the course of their work.

As we continue our examination of the haunting (after)lives of geography at Harvard, we recognize that within this small, albeit powerful corner of the Ivy League, the closure of the Geography Program remains "unfinished business." This wound has been open since 1948, resulting in continual review of the situation by varying institutions. Its afterlife does work: from death understood as peculiarly untimely by those who lived at the center of the conflict, to new, geography-related work beyond the 1950s.

DERWENT'S DEATH

A discussion of haunting must begin with Derwent's death. As we have noted, he was not the only gay man to die on Harvard's campus during this time. We have always maintained that the death of geography and the erasure of this history were intimately entwined with the death of Derwent Whittlesey. The incredible strain placed uniquely on the heart and shoulders of Derwent in the time leading up to and following the closure of the department must have placed unbearable stress on him. These events hinged on the intertwining of his personal and professional lives, despite his best efforts to simultaneously live and hide his love and life in The Loft with Harold and Edward as queer family.

Although both Harold and Derwent retreated separately, at different times, to Emeline's home for sabbaticals and to recover from the devastating events we have detailed, Derwent returned to an increased workload and an altered, depressing reality of turning away graduate students and explaining what had happened to colleagues and alumni across the country in correspondence. He faced all of this alone, as the lone geographer

Derwent S. Whittlesey, professor of Geography and one of the leading political geographers in the country, died yesterday after a short illness.

Well known as an author, traveler, researcher, and historian, he served as a consultant for the U.S. State, War, and Navy Departments during World War II, and was the author of many books and articles on geographic aspects of war and national power. He was also an expert on Africa, and had made several trips there gathering material. He was working on a book about Africa at the time of his death.

Whittlesey was the only professor of Geography at Harvard. He taught courses in political geography, geography of the Boston region, and geography of Africa.

His political geography course was based on "The changing map pattern of the contemporary political world."

Born in Pecatonica, Ill., in 1890, he attended the University of Chicago, receiving his PhD there in 1920. He taught at Chicago from 1919 to 1928, becoming an associate professor of Geography. In 1928 he came to Harvard and was made a full professor in 1943.

He was a member of the Association of American Geographers, and was President of the Association in 1944 and Honorary President in 1954. He also edited the Association's Annals for 12 years.

Professor Whittlesey leaves a brother and a sister. Funeral arrangements will be announced today.

remaining and instructing on campus. We can only speculate about, but also wish not to underestimate, the toll this must have taken on Derwent's emotional and physical health. Once he returned from his sabbatical in Ohio, he lasted another four years, dying in 1956, only months before his retirement planned for 1957.

A short article published in *The Crimson* on November 26, the day after his death, reports his death. We reproduce this short article, entitled "Geographer Derwent Whittlesey Dies," above.

The archives of members of Harvard faculty are not opened until fifty years after their death. This means that Whittlesey's archives were opened in late 2006. We were drawn to Derwent's files—every last boring memo and note. Derwent was queer, and so are we. We fell in love with him and talked with his ghost during our shared year visiting at Harvard in 2015 and 2016. Administrative records are now opened eighty years after their collection, which means that there could be additional files about geography's demise that Conant left behind.

We began doing research on the lost history of this lost department, but, really, ours was the work of what Divya Tolya-Kelly (2006) calls emotional recovery. It was not *what* but *who* we were recovering: not only a lost field but lost geographers and the love they shared for each other. Geography was the only department ever shut down at the college, and it is long believed that Derwent's sexuality in a time of homophobia and McCarthyism played some part. But these were rumors and speculations.

As we did the research, it seemed Derwent's ghost followed us, guiding our hand and drawing us back into the story and the archive at every turn.

Seeing Derwent's face in early March of 2016 was a turning point. He was a striking man, handsome, and he looked happy. We were able to ascertain that he had been with Harold for over forty years based on multiple sources, especially travel records we found on ancestry.com. The Harvard University archives also included records, such as a twenty-two-page letter on Derwent's travels with Harold in Martinique in 1925.

TALKING WITH GHOSTS: THE WORK OF HAUNTING

As we researched and wrote this book, we repeatedly lived out the surprisingly striking parallels between Whittlesey's work life and our own, nearly a century later. Our working lives have intersected with Derwent's in numerous ways, from the minutiae of daily work life, to the desire to please with hard work, while operating at times from marginalized corners of elite institutions. Whereas Whittlesey and Kemp occupied marginalized positions as queer men, our positionality is somewhat different as queer cisgender and transgender women.

What began as simple intellectual curiosity for us, a small archival project off the side of our desks in relation to our main research programs as political geographers researching borders and human migration, the project slowly grew over time into something more political and politicizing, something more meaningful to us both. We felt literally haunted and compelled back to the archives on campus. One day we were meeting in Alison's office, speaking of our need to return to the archives, to get back to it, and the small, rarely used elevator across the hall suddenly opened with no one on it. We paused, looked, and laughed, taking this as a signal from Derwent's ghost to press on, compelling us to return to the archives to learn the story and tell the story.

How does one talk with ghosts? If silences and voids are the essence of haunting, then how does one observe the disembodied? This paradox lies at the heart of *Let Geography Die*: ultimately, while the empirical data that social science typically uses require observation via perception, oppression via haunting is known, acquired, and felt more in the form of lived experience. In this sense, haunting must be experienced in the form of subjectivity and social location. And given that such haunting defies linear boundaries of space and time, attempting to confine or define it using solely analytical frames may miss key elements that bring the oppression to life in the first place. For the haunted, oppression manifests itself through the words, actions, and individual lives of the everyday. This explains how it was possible that an earlier scholar, for example, had accessed the files but asserted to us in no uncertain terms that he discovered no indications of homophobia.

Haunting guides us in the present to talk with ghosts, as a constellation of circumstance and power draw us to perceive the spectral in the first place. And so it was, as two queer political geographers, that we talked with ghosts in our research on the (after)lives of Derwent and geography at Harvard.

Whittlesey was the last standing geographer to teach a course in political geography at Harvard. He taught this course in 1956, the same year that he died. When Alison arrived on campus in 2009 and taught the first course in political geography (to our knowledge) since Whittlesey's 1956 course, students turned out in droves, forcing the third-year undergraduate course to move into a series of ever-larger lecture halls around campus. The course was initially scheduled to run in a seminar room in the basement. When some 100 students enrolled, it was moved to a lecture hall. These students were cartographers, map studiers, researchers of politics, and geography geeks like us who had found their way into geography clubs in high school. The thirst for knowledge and practice, hunger for geography among students who proved curious—like us—to know why there were no geography classes. Although the research would not begin until 2016, during Alison's second visit and second time teaching political geography on campus, the students' inquisitiveness propelled us forward, along with Derwent's story itself.

When Kira found Derwent's final syllabus and brought a copy to Alison in her office on a week when she was teaching this course again in 2016, we sat with it and gasped. It looked oddly like her own. The course was "Readings in Political Geography (Geography 117)—Environmental Foundations of Political Society." In the syllabus, topics covered included geography as a whole, the character of political geography, the state and earth, the ocean and European imperialism, and the Earth viewed as Heartland, Rimland,

and Island. Evidently, as in our own research, Derwent's work extended from the relationship between space and power to understand the operations of states and borders themselves, as imagined through polities in their strategic, creative production and deployment of geography.

Perhaps most importantly, this research was also meaningful and personal for us as queer geographers who came to identify with Derwent Whittlesey in his struggles on campus and in the marginalization of his own voice from his own history. This project became itself a queer metaphorical space during our time on campus, enabling us to move into conversations about our own queer identities and geographies, as we accumulated and analyzed institutional knowledge.

For Kira, personally, it was closely tied with coming out of the closet and transitioning, at Harvard, as a transgender woman. In a time and place rife with transphobic oppression, Derwent's life influenced her own by providing contextual understanding to the meaning of how it felt to be queer yet obfuscated. Though not a cisgender man, she found spectral parallels in being caught in silken symmetries of something bigger than herself: a system prescribing gender and sexuality by virtue of assignment at birth. Like Derwent, Kira had been assigned male and yet, by virtue of her outward expression, coded as feminine her entire life. She later understood this connection from the perspective of trans theory (e.g., Serano 2016; Benavente and Gill-Peterson 2019), where the force of *transmisogyny* could apply to anyone assigned male at birth by virtue of how society treated them for their gender expression. For her, Derwent provided an unforgettable example of how social norms and mores on sexuality were deeply contingent upon understandings of gender itself.

Alison also found that our shared research influenced her and how she identified—more outwardly, more vocally—as queer on campus as a result of this research. There was one lunch with a new doctoral student who came down from Ontario to visit us on campus, what we have now come to think of as "the lunch where everything (and everyone) came out." Alison came to recognize the importance of queer representation and visibility on faculty and how important it might be to share this identity, when appropriate, with students. These are highly relevant to discussions that link this history of institutional homophobia to the present, not only in terms of the absence of geography at Harvard but the presence of queer scholars in visible positions at a moment when sexual identities and practices of identification are being repressed in public and private educational settings and in legislation across the United States. That our collective "coming outs" took place off campus reflected this essence of haunting on campus.

CONSTRUCTING KNOWLEDGE AND QUEER ARCHIVES

Gender and homophobia haunt this story from beginning to end but are not always easy to demonstrate by the standards of empirical evidence to which social scientists hold one another accountable. We encountered more subtle, everyday forms of homophobia as we wrote and sought to publish this story. As one example, we were asked repeatedly by reviewers and avid listeners whether we were sure that Whittlesey and Kemp were really a couple and whether we were sure that Derwent was even gay. We wondered why we had to work so hard to prove same-sex partnership, when a similar relationship or long-term partnership between gender-normative members of the opposite sex might have simply been asserted to be believed.

A persistent issue we discussed, with others, too, was how to locate evidence of homophobia all these years later. One interested Harvard faculty member had gone to Whittlesey's archives and secured permission to access them early, only to determine that no evidence of homophobia existed. We have struggled to reconcile these contrasting interpretations of the archives. To us, there is no question that homophobia shaped not only Whittlesey, Kemp, and Ackerman's lives but the professional life of geography. Homophobia and heteronormative masculinities have also shaped the telling of this history. This prompted us to wonder what, exactly, had been sought, and what might have been missed, given that we were able to find this evidence in these same archives.

Queer archives have become an important topic of interdisciplinary scholarship by historians of sexuality in recent years. Gieseking (2015) emphasizes the intersection of queer positionalities within the archives, and Ahmed (2021) situates university archives as sites of queer institutional complaint. Scholars have asked key questions about what constitutes queer archives and what role is played by queer scholarship. They grapple with the kinds of silences and information between the lines of institutional power and social forces that shaped our own reading of this institutional history through the lens of queer archives.

We agree with feminist scholarship's tenet that positionality matters (Haraway 1991). Feminist scientists and philosophers have long argued that dominant knowledge practices disadvantage socially marginalized groups, due to internal exclusion, denial of authority on knowledge, denigration of "feminine" and "other" cognitive styles and knowledges, use of oppressive theories, obfuscation of these groups, and focus of knowledge production for privileged groups. These disadvantages lead to flawed conceptions of knowledge, knowers, objectivity, and even scientific methodology itself (Anderson 2015).

Donna Haraway (1991) famously reflects on the perspective of the subject through the idea of the *situated knower*, to craft her ideas about situated knowledge. In her view, knowledge itself is dependent on relationships and webs of social connections. This ongoing process means that the social *location* of the knower, informed by identities, social roles, and relationships, deeply affects what a given person can know and how knowledge is produced collectively.

Feminist analysis shows that dominant models of the world can haunt as straight and masculine (Bordo 1987; Young 1990). Gendered relationships may enact differential access to information, like our attempts to access Bowman's archives, or authority on such information, as in our story's previous accounts (Code 1991; Fricker 2007). The projection of gender and sexuality upon skills, cognitive styles, and worldviews further affects some approaches and theories more than others, thereby producing a differential politics of acceptance, as we encountered both in our attempts to publish this story as well as in light of the men who previously tried to delineate its contours.

At the core, then, telling the story of Derwent's life and that of geography at Harvard is a project deeply entangled in notions of gender and sexuality, one that enacts sexist and homophobic norms of conversation and authority of knowledge tied to masculinity (Nelson 1993), one that is part of collective approaches to queer archives.

ON MASCULINITY: ON EFFORTS TO "OWN" THE STORY OF DERWENT AND GEOGRAPHY

We organized this book with chapters that explained the positions, actions, and context of its key figures: Derwent Whittlesey and Harold Kemp,

Emeline McSweeney, Isaiah Bowman, and James B. Conant. But the story did not end with their involvement, nor with their deaths, nor with Ed Ackerman's departure from Harvard and The Loft, nor even with the end of geography itself. In the time since, generations of geographers have tried to explain this history, or to own it, laying claim to their own interpretations. We now must include ourselves in this group.

Across these histories, gender haunts and ties together this story, as another generation of men endeavored to lay claim, to own these histories, to mark their territory. These histories are so laden with masculinities that haunt through attempted ownership of geography's history, its masters and master narratives of what happened behind the gates and along the edges of campus. These hierarchies of power relations, in themselves and the knowledge they produced, provided substance to our haunting via this history before we were even aware of it; the ghosts left in their wake drew us closer to them, and so we sought out their stories.

The homophobic grounds and disciplinary territoriality and infighting that Derwent contended with still exists, rendered in this analysis as fights over intellectual property; they continue to haunt us. Who owns a story? Who owns a narrative? Who has the right to tell the story? Whose knowledge counts, whose voice? Who and what is erased? In our rendition, our hope is that Emeline's voice subverts. We have chosen strategically to foreground Whittlesey's relationship with Kemp and his friendship with McSweeney, where these have been actively dismissed and erased by previous authors.

Whittlesey's relationships with women were also significant, such as his intellectual connection and friendship with Ellen Churchill Semple, or his most intimate relationship with McSweeney. McSweeney, for instance, gave Whittlesey mentorship, comradery, and clarity through the most difficult periods of his life. In a society where heteronormative frames tend to reduce different sex relations to sexuality and romance, the platonic intimacies of women in Whittlesey's history show a wider current in the ocean of his life. In a history heavily narrated by heterosexual men, Whittlesey found refuge in both the queer people and women of his world—of which McSweeney was notably both.

We have worked to complicate the straight institutional history we once assembled. In that history, as published in the *Annals of the American Association of Geographers* (Mountz and Williams 2023), we intentionally designed a paper that we honed to fit the requirements of straight social science narration in human geography. But earlier, and yet in a parallel way, we authored a spectral, queer telling of the same story as a chapter in the book *A Place More Void* (Mountz and Williams 2021). *Let Geography Die* is not simply a synthesis of these perspectives but one that transcends them both through understanding how the stories of Derwent Whittlesey and geography at Harvard lay beyond simple, dialectical representations of the material/objective/straight versus the spectral/subjective/queer. The point is *not* that dualities are absent in historiography but, rather, *also* result from the production of these same dualities. We too are haunted by them; insofar as *we* deploy them, *they* deploy us.

Neil Smith tells the story in such a way as to diminish the role of Conant, emphasize Bowman, and degrade Whittlesey. In his analysis, Whittlesey's masculinity was ultimately insufficient to the task, particularly in that he "had not been aggressive in making allies, either in the administration or among other prominent faculty members" (Smith 1987, 167). Smith adds to this emasculation, writing, "And when the fateful decision came, he seemed wholly incapacitated; rather than fight the decree, he seems to have been resigned to it."

But what if Derwent's actions to fight the death of geography were both hidden and feminized? From his archives, we know that he had forged many strong connections with prominent colleagues and former students. Whittlesey activated these extensive networks, drawing upon these connections in ways evident in his correspondence. And yet his actions were too readily dismissed all these years later, still remembered in the present as weak and insufficient. Whittlesey was emasculated with language such as "incapacitated," the actions he did take feminized and suppressed, until now.

While Smith admits that homophobia *may* have factored into Harvard's decision to let geography die, he uses the presence of economist John Maynard Keynes in American academia as a strawman to attempt to negate homophobia itself, paying no real attention to its role in the entire affair. This, in itself, is a form of homophobic erasure. Whittlesey's alleged weakness of character is repeated years later in a subsequent history of geography in the Ivy League authored by Richard Wright and Natalie Koch. They, too, repeat that Whittlesey was not up to the task of defending the program: "Whittlesey, the only tenured human geographer, proved to be a poor advocate for the discipline. . . . Yet Bowman's lack of support probably sealed the program's fate" (Wright and Koch 2009, 619).

Martin's "invited response" to Neil Smith's essay proved distressing to read. He aggressively erases the romantic relationship between Harold and Derwent (what, today, we would call partnership), by referring to both repeatedly as "close friends" (Martin 1988, 152). He also attacks Whittlesey's character most directly, by writing of Whittlesey's, for instance: "And since Harvard's strength in geography had . . . been in physical geography, it might have been wiser for Whittlesey to build at least somewhat on tradition. . . . But Whittlesey's insistence on bringing his long-time friend Kemp into the department as instructor (1930) was beyond the injudicious. It was a mistake" (153).

Doubling down on this attack on Derwent's masculinity via his relationship with Harold, Martin adds, "Yet the fact remains that Whittlesey had not developed a meaningful geography program, had not made his mark on the Harvard Yard, and above all had selected a departmental colleague—Kemp—on the grounds of friendship rather than capability." And despite his commentary's antagonistic tone toward Smith on Bowman, Martin mostly agrees with Smith's assessments.

We find much collective discussion and dismissal of Derwent and his perceived failures, when the focus might have been turned more fully on discussion of Conant and *his* actions, and on collective institutional failures and exclusions. As this history has been memorialized, Derwent himself was dismissed, erased from, and blamed for the story.

Beyond restoring him to the story and humanizing Derwent, we grew enamored with Derwent, admittedly, because he was smart, funny, and kind. He was successful, excellent at the work that he did in research, writing, teaching, and advising students, even if he is not always remembered for these accomplishments.

HOW THIS PROGRAM'S HISTORY HAUNTS THE DISCIPLINE'S PRESENT

> Whether we speak of the elimination of Geography at Harvard nearly four decades ago (1947–51), or in recent years at Michigan, Northwestern, Chicago and now Columbia, many of the issues raised in the course of events at Harvard were repeated in different forms at these other universities. We must be mindful of these experiences, as we seek ways of strengthening the position of Geography in American universities. (Cohen 1988, 148)

This history embeds itself into legends of geography's (after)life at and beyond Harvard, such as its loss at other universities, like the University

Figure 7.1
Stairs leading up to Harvard's Secret Garden. Photograph by Rowan Flad.

Figure 7.2
Empty bench in Harvard's Secret Garden, once a popular cruising spot for queer
people since at least the 1920s. Photograph by Rowan Flad.

of Michigan and the University of Chicago, and within the Ivy League
(Wright and Koch 2009). We can also trace the (after)lives of geography at
Harvard through recurrences and hauntings on campus at Harvard from
1960 to the present, ending with our own arrival on campus as political
geographers in the 2010s.

This story has been told slowly, by geographers, and by us, for so many
reasons. It is a story we puzzled over in writing and in conversation, seated
on the benches of Harvard's secret garden, hidden along the edges of its
campus and pictured in figures 7.1 and 7.2. These absences left us seeking
clues in the archives and residual material infrastructure of geography left
behind on campus.

Political geography also has its conversations with ghosts. There are, for
instance, ghosts of imperialism wherein maps became projects of empires
seeking to expand their territories. There are, further, the ghosts of racism
and environmental determinism in its early studies of the correspondence

between regional climates of the world, which were mapped onto "races" of people. The ghosts of geopolitics itself and its early theorists such as Ratzel, whose argument that nation-states were like living, breathing organisms that needed room to grow were adopted by the Nazi party, for example, to justify its occupations and genocidal policies. Because of these histories, political geography would diminish and fall out of favor, only rising to prominence again by the 1990s and resurgent again after 9/11 (e.g., Agnew 1994; O'Tuathail 1996).

As students and instructors of political geography, the ghosts of Derwent Whittlesey and geography haunt us, influence our actions, and led us to create this book. These ghosts of disciplinary histories can never be divorced from the ghosts of their leading theorists and practitioners. This was true not only of Derwent Whittlesey but of many geographers who sit on our shoulders as we write, looking on. If we could have a conversation with Neil Smith now, we would ask about how he formed his analysis and share stories about how we formed our own. We would talk about queer archives, and about whether they can only be assembled by queer people. Any telling must acknowledge the silences between the lines and around the edges.

ON SECRETS AND AFTERLIVES

And so, we found ourselves sweating on a hazy, humid day during that last week of June in the City of Cambridge's court records at the Middlesex Probate and Family Court. We were lost in a bureaucratic maze: bewildering shelves upon shelves of records of obituaries, hand-recorded in monochrome binders that had been damaged over the years by flood and fire, adhering to an internal logic not rendered quite knowable to outsiders. We were redirected by city workers from one room to another, one floor to another, to comb through a byzantine swirl of files. We were nearly undone in the end, when finally directed to a copy machine minutes before the building closed to make copies of the wills we were so delighted to discover, only to learn that we needed exact change to make it work, without instructions on how to operate the machine.

We were, in that precise moment, ensnared not only in our own adrenaline and sweat but in our then nascent pursuit to better understand the entanglements of two deaths: the study of geography at Harvard and Whittlesey himself. We knew that we needed to learn more about the intimacies of both. Throughout our research we had sought empirical evidence to lay to rest elements of this story that remained uncertain, including, for example, details and confirmation of the romantic relationship between Whittlesey and Kemp. The archives left open no question that the two had been together for decades, from their correspondence to travel records and shared residence. We went in search of their obituaries to learn more about their deaths and, specifically, the guarantor of their wills. This guarantor turned out to be geographer, housemate, and confidant Edward A. Ackerman in both cases. These wills and their recording registered by Ackerman in the City of Cambridge add to the material traces of geography and geographers at Harvard. The discovery of Ackerman as guarantor was not shocking to us, but a sweet feeling of relief accompanied the realization that we had correctly understood his importance in the lives of Derwent and Harold, still, to the end. After all those years, long after he had lost his job at Harvard and moved geographically away from this past—he remained an important and intimate figure in their lives and, ultimately, in their deaths.

Correspondence between Whittlesey and Ackerman substantiates that Ackerman's termination by the Conant administration in 1947 was both a personal and professional blow to all three men and the family life that they had built together at The Loft. We eventually learned from Edward Ackerman's archives that he maintained his emotional closeness to Derwent and Harold.

QUEER FAMILY AT THE LOFT, INTERRUPTED

Our protagonists each struggled in their own way with the unfinished business of geography in the immediate aftermath of its demise. Ackerman threw himself back into his work, as was his habit, while Whittlesey struggled to do so when he finally returned to campus. As always, they took to correspondence in search of connection and healing, traversing the distance with words of love.

Toward the end of our archival research, we located—in Ackerman's archives—an emotionally rewarding and comforting exchange between Ackerman and Whittlesey, written between autumn of 1948 and spring of 1949, following Ackerman's departure from Harvard, Cambridge, and his shared home in The Loft. At first, in August, Ackerman sent his resignation letter to Whittlesey, with a handwritten cover note on University of Illinois (Urbana) letterhead stating that this is the only copy of his resignation letter he has, "In case you need to refer to it—."[1]

Following his resignation, Ackerman departs the United States and returns to his work in Japan. Whittlesey writes him there: "You are much in my thoughts, as you must guess. Not merely because the college is opening, and I miss you there, as well as at home, but also because you are as much a member of the family as though you were in your little room."[2]

Whittlesey also writes to Ackerman about the pain and discomfort of his own return to campus in September, in the aftermath of the administration's decision and Ackerman's departure: "Today is the first day of classes, and I am thankful to be kept exceedingly busy. I started out to have a last bit of vacation, partly because I realized that these beginning days would be difficult. How difficult, my imagination did not forewarn me."[3]

Meanwhile, Ackerman writes to Whittlesey from Tokyo, at a time when all three men are clearly reeling from their recent and distant separation from their shared home of ten years in Cambridge: "I should have written you long since, for not a day passes without you entering my thoughts."[4] Ackerman details Harvard's underhanded insistence that he return quickly without informing him that he would be doing so at his own expense for a reduced workload and rate of pay of which he was uninformed. Ackerman holds Dean Buck accountable: "I intend to part ways with Mr. Buck quietly, but only if he is decent."[5]

Like Whittlesey, Ackerman is grateful to turn to his work to cope with the departure from Harvard: "I am glad to have had this interest to fall back on at this time, for the parting from Harvard would have been much more difficult without work here. It has helped me to maintain confidence in the validity of my approach, and in my fitness for dealing in the world of ideas. My love, as always, Ed."[6]

Their correspondence clearly continues, much of it inaccessible to us. It is clear that Ackerman returned to the United States from Japan. Whittlesey writes to him, stating:

> The registered letter and package arrived from Chicago. You needn't have sent the wrist watch back. Since you did, I confess that it is more convenient than the pocket watch I have been using. Yes, I knew the crystal was cracked; I am sorry to have caused you any qualms—if finding the crack did make you wonder how it happened. The powers of attorney will go back into the safety deposit box. The confidence in me you express by that document, the excerpts Harold read me from your letter, and your visits here go far toward making this stage of my life worth living. The sharpest pangs of the past year have come from your absence from home. That it hurts you, too, only shows that you are one of the makers of the loft that has become a home. I am thankful that the pain of separation is enfolded in our unity of spirit. So long as our mutual understanding holds, being in different towns cannot separate us.[7]

Whittlesey continues: "One of my consolations is a feeling that you may profit in the long run from what seems now to be only loss, if being away from home stirs you to create a home for yourself. (I have learned that a blow that has bowled me over can turn out to be a push in a new and right direction)."

He then addresses Ackerman's search for a partner with whom to make a home, using a quite out-of-this-time word—partner:

I know how chancy it is to find the right partner to make such a home. Now that you have waited this long, it would be sad indeed to make a mistake. I trust your judgement of others, but not your evaluation of yourself. "don't" undervalue your very remarkable qualities for making the right kind of home. I needn't name them, because you are intelligent and therefore must know what they are. If I know you as well as I do, you have just about every trait needed except gaiety. When you find the congenial person with enough joie de vivre to carry you in train, life will be made to spark for you forever. Harold has done that for both of us, and without whom I would have become stodgier than I am. Above all, don't hesitate on superficial grounds. The person for you now is likely to be a good deal younger; she might even be richer.

Please forgive unasked advice. I proffer it only because you are part of the family. You have a long life of happiness ahead, if your home life goes well. I shall find much joy in your personal success, no less than in your professional success.

Your doting, Whit

We found these passages comforting proof that queer families, spoken of commonly in the present as "chosen," while less spoken of at all in their time, nonetheless persisted against all odds. Although separated by all that came to pass, the love between Whittlesey, Kemp, and Ackerman persisted until the end.

A FORK IN THE ROAD

Why did it not occur to us to take a photograph of ourselves on that momentous day that we went to the city archives together, shown in figure 8.1, one of our last that year in Cambridge? Maybe because it was swelteringly hot, that early humid, oppressive heat that surprises with its first arrival in June. Maybe because somatic memories run deep. Kira had just begun dressing as a woman at work, and this was one of our first forays out into the world together, as working women. We had no idea what these new (to us) archives held in store for us, but it was a day of adventures, well worth the sweat.

We wish we had photographed ourselves that day to see all that we remember. In April of 2017, ten months later, we returned to Cambridge

Figure 8.1
The City of Cambridge archives. Photograph by Alison Mountz.

to give three presentations about this archival research for the annual meeting of the American Association of Geographers (AAG). The first was at Harvard's Weatherhead Center, upstairs from the Center for Geographic Analysis, part of the AAG Political Geography preconference. The second was a walking tour—shown in figure 8.2—attended by about thirty conference-goers. On a cold, gray day, we led our group, comprising primarily geographers, through Harvard Yard, past The Loft, and Agassiz Museum where the new Geography Program had been slated to be housed in the 1940s. We ended inside the large lecture hall (shown in figure 4.5). This is where we explained the climax of our story, the fateful notes that Bowman had taken about Conant's private conversations with new members of the Board of Overseers the night before the board met and took the vote to affirm closure of the department.

We were glad to be warm inside the lecture hall, if chilled to sit in the very room once occupied by Derwent, Harold, Edward, and their students. We sat as geographers in what had been purpose-built as the Geography

Figure 8.2
The authors lead a walking tour about the history of geography on campus, beginning in Harvard Square before entering the gate onto campus. Photograph by Rowan Flad.

Building, once housing Rice's Institute for Geographical Exploration.[8] We found ourselves nestled not only at this fold in history but at a historical-spatial intersection: a fork in the road leading to two symbolic locations, each representing a divergent path to a distinct future. In one direction lay The Loft, across the street from President Conant's office and home, the Faculty Club where Bowman stayed from time to time when he came to town, and that sunlit meeting room where the board cast their final vote. In the other direction lay geography's once promising and promised future: a beautiful new home in what was then called the Agassiz Museum, decades later (in 2020) renamed as reparation for Agassiz's advancement of racial segregation and white supremacy through science (Harvard 2023). The building now houses the Department of Anthropology and the Peabody Museum.

Our third presentation was at the AAG conference itself in downtown Boston, a presentation in a series of sessions aptly titled "Into the Void" (Kingsbury and Secor 2021).

When is a death not really a death? There we all were, after all, quite alive. More than 9,000 people had traveled from around the globe to attend the annual geography conference. Thirty-two of us stood on this threshold to relive this history.

Here, we recounted stories of some of our own efforts to tell this story, including contacting its living representatives, like Whittlesey's students and relatives.

THE LOFT AND THE ARCHIVE AS CLOSET

The contested history of the erasure of geography from Harvard's campus and curriculum is important because it is still taught, repeated, recorded, and recounted incorrectly in graduate programs where students learn geographic thought, theory, and history. The story is passed down in a manner that inadvertently recycles the homophobic narrative about Whittlesey as a weak character blamed for geography's closure, the "wrong man" to save it because he could never be "the right man for the job." Geographers lament this loss frequently and publicly. But these lamentations do not realize just how closely the life and death of the department were tied to Whittlesey's life—personal and professional—and death, from his hiring to his demise as the final geographer teaching on campus in 1956.

One of our favorite things that we read about The Loft is that geography students and faculty gathered there, on a regular basis, for drink, for discussion, for togetherness. This made the place feel familiar, accepting. We could picture it and imagine how much fun it would be to gather there. We need these spaces as students, as colleagues, as queer people, and we remember them well years, and even decades, later. We may not recall the exact conversations we had but remember laughing with women we would gather with as graduate students and as new faculty members, struggling in those early years. We can well understand how important The Loft would have been, not only to its three inhabitants but to the regular visitors— queer geographers like us who would have found refuge and camaraderie there.

There is no question that The Loft and the archive were queer spaces, then and now: radical spaces of challenging, subverting, documenting. They were important spaces of knowing and telling. Sara Ahmed (2021) identifies the importance of bureaucratic processes within universities as sites of institutional complaint. As she notes, although little may happen in the time of recording, this documentation is nonetheless an important political act for its radical potential to tell as it leaks from the archives into the future. It is precisely this reckoning with the past that compelled us to tell these stories in the present.

Our own queer revelations and struggles with gender, sexuality, transphobia, and sexism along the way, in real time, placed us in ever more intimate proximity with the content of the story unraveling—our comings out as we ventured into the quiet, stormy solitude of institutional archives. Through our stories, our lives themselves became intertwined in their own way with that of Derwent's and geography at Harvard; simply put, the research *changed* the researchers.

> Haunting raises specters, and it alters the experience of being in linear time, alters the way we normally separate and sequence the past, the present and the future . . . Haunting and the appearance of specters or ghosts is one way . . . we're notified that what's been suppressed or concealed is very much alive and present, messing or interfering precisely with those always incomplete forms of containment and repression ceaselessly directed towards us. (Gordon 2011, 2)

Doing this research also changed how we relate to and understand each other. At some point, we casually—yet seriously—granted one another

Figure 8.3
The entrance to Harvard University Archives. Photograph by Alison Mountz.

passage and trust to care for our own office files and archives in the event of our own passing. For, throughout our engagement with Whittlesey's archives, we also came to understand just how important the records of own lives—of anyone's life—might be to those who would come after us, perhaps asking questions of us as we have of Derwent's ghost. We understand just how many stories the archives might offer up, the care they might require, and the trust between us cemented by our years-long collaboration.

Archives blur public and private lives and spaces. They reveal such personal intimacies and yet are housed to be made accessible to publics, with that access rendered conditional in ways that police and regulate identities through forms of documentation and stature that a person must provide to access archives and the sorts of behaviors, rules, and regulations they must conform to within them. Feminist scholars like Dolores Hayden (1997) and Gillian Rose (1993) have long challenged easy divisions of public and private, noting that much scholarship overlooked the inherently political to be engaged within private spaces (Marston 2000), like the domestic sphere. As Mona Domosh (2007) shows in her work, archives offer important entry points not only into forms of documentation and complaint that were not allowed in their time, as Ahmed (2021) argues, but also a medium to engage and foreground the political in private spaces and smaller scales of queer homes, families, and relations historically excluded and overlooked as part of the political. The private lives of Whittlesey, Kemp, and Ackerman were made public by force; they had little choice in this matter, even as they remained closeted and hidden in The Loft and the archive all this time.

Furthermore, whereas Conant and Bowman carried the privilege of curating their own archives for future public consumption, we recognize that Whittlesey and McSweeney enjoyed no such privilege or opportunity. At the same time that we have reveled and rejoiced in the chance to read and learn from their intimate correspondence in the archives, satisfying our intense desire to hear and document their voices, we also recognize that doing so simultaneously reproduces hierarches of power, performance, and institutional rules designed to govern power and stature.

For these reasons, we are drawn time and again to revisit the social and political locations of archive and loft as sites that signal so many things at once: being closeted, being protected, having voice, having no voice, having future voice. These are sites that blur not only public and private spaces but professional and personal lives, acutely so in the case of Whittlesey's and Bowman's files. The location of archive and loft as closet is at once safe and liberatory, and radical and oppressive, and also eventually both public and political.

These are sites that convey politics and power in where they are held and kept, how they end up where they do, who can access them, when, and under what conditions.

These are sites that harbor secrets.

What would Derwent have purged from his archives had he had the opportunity? What would he have considered too private for future readers in Harvard's archives? What—and who—might we never have known had he had this opportunity?

HAUNTING GEOGRAPHIES, HAUNTING GEOGRAPHERS

As a loyal alumnus of Harvard, I have sought to convince myself that the geography decision was for the best interest of the University. I must admit that it has been impossible for me to do so. Careful listening and much thought have only left me profoundly disappointed in the University's action, both from an intellectual and an ethical point of view. That is not because of the inconvenience which the decision may have caused me, but because the procedure and justification were disturbing in the implications they carried about American educational administration.[9]

These premonitory lines were handwritten by Edward Ackerman in his one-page letter of resignation to Dean Buck in August of 1948. We have reviewed the history of geography at Harvard, enhancing previous work by offering a fuller, more complicated story grounded in recently released archival data. Although previous accounts of this history were important, we provide new evidence suggesting that the impact of homophobia on the Geography Program's closure was stronger than surmised. As we have shown, the feminization of both human geography and queer geographers on Harvard's staff was responsible for direct opposition of the Conant administration, which culminated in a policy to let geography die.

It is worth reflecting briefly on what the history we recount means for the discipline of geography, geographers in the United States, and for people associated with Harvard: students, faculty, alumni. The history of geography at Harvard continues to haunt, having not been dealt with, and, therefore, offers crucial lessons for the discipline. Much has been made of the wider reverberations of this loss of geography at Harvard. The closure of Harvard's geography department still haunts geographers—a haunting articulated earlier in the words of the late political geographer (and student of Whittlesey) Saul Cohen. Many believe that this closure emboldened administrators across the country who did not well understand the discipline to close programs in economically challenging times of consolidation. Eventually, the University of Chicago (where Derwent was previously employed) closed its department, as did Yale and Columbia, leaving Dartmouth College as the sole university in the Ivy League with a geography department (Wright and Koch 2009). Indeed, Ackerman and Whittlesey themselves predicted these consequences often in writing, including in Ackerman's handwritten letter of resignation to Dean Buck with which we opened this section.

One month after we ventured into the City of Cambridge archives, in July of 2016, in his monthly column to several thousand members of the organization, geographer Glen MacDonald (2016), then president of the American Association of Geographers, happened to make passing reference to the collective knowledge of the end of geography at Harvard. Such casual references are common, even though geographers know this history only sparsely—or more accurately, we collectively misremember—a collective mythology forged through oral history, passed along and canonized in introductory graduate courses in geography.

One last lesson of this history is that geographers and geography as a field tend to flourish or decline together, whereas conflict between subdisciplines and feminization of particular strains of knowledge and those who pursue them harms geography as a whole. It thus tends to be the case that the strongest, longest-lasting geography departments make space to support a variety of methodologies and collective works—both theoretical and applied. While somewhat anecdotally based in our own careers and observations across North America, this observation bears out and bears repeating.

The discipline is always strengthened not only by a diversity of knowledge forms and methodological approaches but by its diversity of people. The exclusion of queer geographers and geographies remains inexorably tied to geography's erasure. We need to understand this exclusion because any field is shaped as much by its absences and silences as by what—and who—is present. Our own unlikely presence as geographers at Harvard fueled our curiosity, serving as initial provocation for research into geography's spectral afterlives. We feel connected to Derwent Whittlesey and haunted by this history. We were struck in reading the archives by how much our daily work resembled his. We focused on extensive empirical documentation of a history suppressed for too long, noting our own queer positionality within the in/stability of the queer archive (Gieseking 2015).

We were struck, in reading an article about geography published by *The Crimson* in 2016, by its absences and omissions, by the erasure of queer people and homophobia from the story, including the archival history we had shared with a student journalist who wrote a feature-length story for the paper only months prior and which was never published.

Our telling builds on what Natalie Oswin (2020) calls "an other geography," in which—in the face of continued marginalization of others—solidarities form among these others. We understand ourselves and this book to be part of both the marginalization of queer people and knowledge and part of the recovery of silenced histories, lives, and contributions. As Sarah Ahmed (2021) argues, documenting institutional oppression is itself a queer form of complaint, a queer institutional map. Through Ahmed's work, we understand Harvard's archives as institutional closet where the documentation of a complaint, even if not attended to or helpful to people at the time of its filing, becomes documented. This project was also itself a queer metaphorical space for us during our time on campus, enabling us to move queer geographies and institutional knowledge from margin to center of our research.

Elements of this history also speak powerfully to the contemporary state of the university and its ongoing neoliberalization, one where private interests matter even more extensively on campuses than they did in the time when Derwent and Harold fought together to preserve geography at Harvard. Money talks, as they say, and it always has.

Over time, for example, projects like the Institute for Geographical Exploration, always subject to politics and their struggles, have become more common, with donors influencing which programs and buildings emerge on campuses; however, such conflicts need not end entire programs, as they did with geography at Harvard.

A RETURN TO THE BEGINNING: A SERIES OF INCONSISTENT
EMPIRICAL FACTS

There are different ways to read the empirical evidence set forth in this book. This is a story about a particular place and time in the history of the discipline, told here with a focus on key figures. However, it is not an isolated history but, rather, one that reverberates with other, well-documented institutional histories that turned on the exclusion of queer people (e.g., Schnur 1997, Johnson 2006). We remain committed to restoring Whittlesey's perspectives, as we understand them through his archives, to the historical record. Ultimately, institutional and societal structures maligned and oppressed Whittlesey, despite his tireless efforts to advance research and instruction of a discipline he loved.

As we have shown, Derwent has been memorialized as "weak" in disciplinary histories. But his characterization as "poor advocate" (e.g., Wright and Koch 2009) contradicts an abundance of evidence in the archives demonstrating precisely the opposite. We are painfully aware that we have not focused amply on his scholarship and the recognition garnered by peers, including the editorship at the *Annals* and presidency of the AAG. Nor have we discussed as fully as we might have the content of his classes, the caring mentorship and loving friendships on display through his correspondence, or the content of his scholarly contributions. These will be the subject of future writing by other scholars who we hope will join us in these collective queer projects of restoration. By analyzing archives made available more recently, we have sought to restore Whittlesey's struggle to keep geography alive to the historical records. In so doing, we extend, and at times correct, an institutional history begun by Morris and Smith.

When we started this archival research, pursuing evidence and threads of conversation begun by Neil Smith in his 1987 essay, we encountered a number of empirical facts in the archives that proved inconsistent with the way that this history is often remembered and shared. People suggest, in writing, that the field was in decline and that Whittlesey was not a strong enough character to save it. We have been able to prove empirically that neither of these assertions was true, even though they were circulated repeatedly by more than one administration at Harvard. These assertions were advanced as rationale to support the end of research and instruction in geography, including refusal to respond affirmatively to any of the five subsequent reviews recommending that this decision be reversed and the program be reinstated and expanded.

We have provided extensive empirical evidence to disprove a number of false empirical assertions, summarized briefly here. It was not true that enrollments in geography courses or graduate instruction were in decline.

On the contrary, they were increasing even as the administration starved geographers of resources.

While it is always the case that a key function of university administrators is to make difficult budgetary decisions amid constraints, it was *not* true that the university was facing a moment of budgetary crisis or decline when the decision was made to let geography die. On the contrary, both Harvard *and* geography at Harvard were thriving, measured empirically by finances and enrollments. It is also true that these were entwined and that geography was finding success institutionally: training men for "war work" and attracting veterans and other students once the war ended.

It was also not true that geography was a subject that Harvard could just not do well. On the contrary, a small number of geography faculty, led by Derwent Whittlesey, had quickly made the department a leading program nationally, evident in correspondence between prospective students and faculty members seeking to come to Harvard, in the national success of the professional and scholarly reputation of faculty members in geography, and in correspondence between Whittlesey and colleagues all over the country. As Whittlesey himself wrote in a letter to geography alumnus Charles Leonard in 1951, "After the war, there was an enormous increase in interest among both undergraduates and graduates, and we had a first class small department. Now I am the only one left."[10]

Lastly, it was never the case that Derwent Whittlesey was weak, a poor advocate, or not up to the task to save the program. While his masculinity was very much tied to the end of geography, he should not be blamed as weak or the wrong character to lead.

While arguing that we have disproven these standpoints, we also recognize that this story is complicated. Institutional histories and politics might always have moved in other directions. Certainly, for example, Rice's Institute could have continued to provide infrastructure for geographical instruction, research, and material resources at Harvard, as geographer Dudley Stamp argued in his institutional review in 1952. Similarly, the contemporary Center for Geographical Analysis could prove foundational for development and restoration of geography at Harvard.

People who lived this history clearly knew some or possibly all of these facts. Our strategy here has been to cast them alongside each other and discussion of queer positionality, queer spaces and archives, and our efforts to locate homophobia therein.

We also recognize that positionality in this history is fraught, shaped by the whiteness and class privilege of a rarefied group of exclusively white men. Women appear only fleetingly in this story in caring correspondence and in passing as instructors, and people of color appear not at all—even though Black Americans proved central to the building of the Ivy League (Wilder 2013). White masculinity proved central to this story: what kinds were and were not allowed, and what kinds were exploited, demonized, and blamed. Ultimately, Whittlesey, Kemp, and Ackerman found passage to instruction at Harvard as white men. But they were never accepted as men attracted to other men. While they attempted to hide their private lives in The Loft, they nonetheless made their way into university archives as queer people. As such, both The Loft and archive serve the many functionalities of the institutional closet: a space at once oppressive in its silencing and liberating in its communality for those who shared its space and discoveries. The Loft was a popular space for faculty and students, an escape from oppressive forces operating against individuals and geography more broadly. The archive enabled us to discover and document this history.

Existing institutional histories suggest that geography's story at Harvard ended in 1948, following Ackerman's termination due to Conant's decision to reject his case for tenure. But along the way, we have provided new, distinct empirical evidence as reminders that there is more to the story than collective memorialization might have suggested. Three committees and one university president voted affirmatively on Ackerman's tenure and promotion case. Not one, nor two, nor three, nor four, but five institutional reviews at Harvard and beyond Harvard recommended consistently and repeatedly that Conant's decision to end geography be overturned. Two men fell in love and shared a life together for forty-three years. Three men lived together as a family in The Loft for ten years, queer by association. One university president defied all previous recommendations by deciding to turn down one of these men for tenure.

New evidence we have introduced also shows how geography's story—including the contestation of this closure—continued into the late 1950s and recurred in small spurts across the university over the past sixty years.

Geography's decline, then, was slower than Morris (1962), Smith (1987), or Martin (1988) understood, faced more resistance than existing accounts identify, and is more incomplete than the way it is popularly remembered in shared mythology.

Contrary to the popular notion of its abrupt death, geography lived on at Harvard after its formal decline as a subject of undergraduate and graduate education and research, though its times and places of instruction and research were few and far between, and geography today is accorded little status or understanding at the university, beyond the important work of the Center for Geographical Analysis. The Graduate School of Design (GSD) brought Associate Professor George Lewis as a visiting scholar for regional planning and geography in 1960 (Harvard University 1961). He would later teach geography as an extension course in 1967 and 1972 until 1975 (Harvard University 1968; 1973–1976). The GSD was a center for spatial analysis and "theoretical geography" throughout the 1970s. Examples of its geography-related activities included conferences, exhibitions, and laboratories (Harvard University 1970; 1976). It also received major National Science Foundation grants to produce cartographic data. Harvard established the Center for Geographical Analysis, housed in The Weatherhead Center for International Affairs, in 2005, founded by history professor Peter Bol (CGA 2017). Alison taught courses in political geography on campus in 2010 and 2015 that were not so dissimilar to those offered by Professor Whittlesey.

OUR LOVE FOR DERWENT S. WHITTLESEY

It was hard not to fall in love with Derwent Stainthorpe Whittlesey when reading his correspondence. We fell for Derwent above all because he was charming and thoughtful, taking time and care in his extensive correspondence to colleagues, students, family, and friends around the globe. We loved Derwent for his letters to Emeline in Wooster: those beautiful lines and the vulnerabilities and love each conveyed to the other, not to mention the brilliant, hand-crafted "picker-uppers," which so delighted Emeline that she devoted the content of at least two letters to their description, her gratitude and glee at all that they enabled her to accomplish.

Derwent meticulously filed these letters, and we meticulously devoured them. We loved him because he seems to have always tried to do the right thing. We empathized not only because we too are queer geographers but because his mundane work life was surprisingly like our own. We held in common with him a surprising number of things, working some seventy-five years later as queer political geographers at Harvard. We read and write today about how universities have changed, but Derwent's quotidian life was much like our own. The archives showed squabbles over reimbursement, apologies for late manuscripts, and letters of recommendation for students. Consider how aptly Whittlesey's description of academic life in 1945 applies to academic life in 2025: "Don't I know what fritters away the time of the man who is paid to teach, promoted for being a scholar, and kept busy with administration?"[11]

Derwent Whittlesey's archives and correspondence demonstrate what happened to geography over time, including the loss of Ackerman. They forecast the subsequent haunting of this void across the country.

In a letter to Lt. Colonel Hubert Schenck, who wrote to FAS Dean Buck in support of Ackerman, Whittlesey writes:

> Geography barely got started at Harvard 20 years ago when the long financial depression precluded all further expansion. When the war and postwar demand struck, we found ourselves seriously understaffed. Unlike many universities, Harvard has no[t] increased its faculty because the administration expects to reestablish a limitation on the student body. It so happens that Ackerman is the only young man in the country for whom we can obtain a permanent post at this time, partly because he is favorably known here, but chiefly because he is the top-ranking geograph[er] of his age-group. If he comes back to us in February, we shall be able to go forward. If he does not, the subject will disappear from the Harvard curriculum. That would be a body blow to an important earth science throughout the country, because other universities look to see what Harvard does.[12]

Ackerman himself, in his letter of resignation (quoted earlier), wrote not about his own personal feelings of disappointment and betrayal by the alma mater he loved and the employer to whom he was devoted but of a principled stance against the erasure of geography from American higher education.

Conant's administrative records also reveal the demise of geography in formal reports and informal correspondence. For example, Harvard's administration contracted Professor Dudley Stamp at the London School of Economics to conduct research and write a report on geography. There is correspondence to Stamp from Conant about his (Conant's) inability to meet with Stamp in person, but his provision of notes on the report, nonetheless.

Here Conant and Stamp's views on Whittlesey emerge: "I believe he has allowed himself to be unduly influenced by his friend Kemp—a dilettante whose interests in art and music—admirable in themselves—have led him right away from any serious contribution to his subject for many years" (Stamp 1952). Also indicative of homophobia's quiet haunting of this entire affair are Bowman's meticulous notes in the margins, left behind in his archives after the Board of Overseer's meeting, previously described, where he identifies Whittlesey's character as a chief reason for Conant's policy. Whittlesey's documents included extensive personal correspondence, including letters between Kemp and Whittlesey and long, loving, mischievous letters from McSweeney, who knew everything. By virtue of not "burning the letters," Derwent's surprisingly complete archives became available to us to sift for evidence of his love, his life—and became an active site of his haunting.

Bowman, in his role as new member of the Board of Overseers when Conant shared this decision, also left handwritten notes in the margins and secret files for future historians. Whereas Bowman invested heavily in curation and conditions of access to his archives (restricted to men of a certain age and of international repute), Whittlesey, we believe due to his sudden and untimely death, had little intervention in the organization of his archives—only meticulous organization of files likely moved directly from his office when he died. Bowman wrote:

> Of course, I am surprised at the thoroughgoing way in which geography will be practically eliminated. . . . The declaration that geography is not a university subject I shall attempt to counter on broad lines. . . . Some time when we meet, I would like to talk about some of the more intimate conditions in the geographical field at Harvard. . . . The forces in a university are complex and I suppose there is much on the other side that you and I do not know but which will come out in the course of time.[13]

In Gordon's (2011) analysis, haunting is about oppression that may appear to be over and done with but never actually disappears—it is always present and making itself known, if not seen.

> Something is being freed and there's a reach for it. The reach is key. The something-to-be-done is not ever given in advance, but it can be cultivated towards more just and peaceful ends. This emergent rather than fatalistic conception of haunting often (to the extent that it is or is becoming an explicitly subversive or rebellious consciousness) lends the something-to-be-done a certain retrospective urgency: the something-to-be-done feels as if it has already

been needed or wanted before, perhaps forever, certainly for a long time, and we cannot wait for it any longer. We're haunted, as Herbert Marcuse wrote, by the "historic alternatives" that could have been. (Gordon 2011, 5)

Let Geography Die is our collective "reach" to recover and restore some of what has not been spoken from geographers' collective memory of this history. In our archival research and discussions, we found ourselves early on joking about Derwent's ghost. But as time went on, it became more appropriate to draw on haunting as a framing. Everything in this history comes back to Derwent's ghost and his hauntings of the present. Geography remains unfinished business. We continue to try to make sense of what Derwent left behind for us and how we (geographers, Harvard students) are all part of—haunted by—the story we have only begun here. What can we learn from the history and politics of geography, and from Whittlesey's story?

THE "SOMETHING-TO-BE-DONE"

Which audiences are ready to hear this history? During both of Alison's years as a visiting faculty member at Harvard, students were hungry to learn political geography and eager to learn this history as well. In February of 2016, Alison organized and moderated a panel of geographers to speak on this institutional history and on why geography matters. Panelists included Peter Bol (then vice provost of Harvard), Mona Domosh and Richard Wright (Dartmouth College), Tim Cresswell (then at Northeastern University), and Kira. The event had standing room only. Not long after, a journalist for the student paper, *The Crimson*, wrote an extensive investigative story on the history, twice interviewing the authors of this book. Her essay was never published. Instead, weeks later, in May of 2016, the paper published a much shorter article by a different student journalist, which consulted none of the archives and repeated many of the fallacies sustained by the mythology of this history.

We have to accept some uncertainty in our interpretive analysis, drawing on what Jack Gieseking (2015) calls "useful in/stability" in the unpacking of the social and spatial dialectic and thinking through queer archives. By useful in/stability Gieseking refers to the archive as "both stable in its physical form and unstable in its sociality" (2). For us, useful in/stability proves

helpful in embracing the uncertainty of silences in the archive and in placing correspondence in the context of its time, as we read and imagine what
it would have been like to have been Whittlesey, Kemp, and other queer
faculty and students at Harvard during the Cold War.

Of course, geography is still practiced in some ways at Harvard, our own
presence, visits, and teaching of political geography there notwithstanding.
Although Whittlesey died in 1956, visiting scholars taught courses until
1958. In the 1960s, the lab for spatial analysis was founded, its history documented by geographer Matthew W. Wilson (2017). In the 1970s, this work
took place in the Graduate School of Design. The 2000s saw the founding of the Center for Geographic Analysis and its expansion in the time
since. We believe that 2010 and 2016 were likely the first years that political
geography courses were again taught following Whittlesey's retirement in
1956. In these ways and others, Whittlesey's history is our own. How has
this past (institutional, homophobia, Secret Court, geography's insecurities,
Harvard's void, oral history on campus) never gone away?

> For better or worse, the emphasis on the something-to-be-done was a way
> of focusing on the cultural requirements or dimensions of individual, social,
> or political movement and change. And one of those requirements was that
> the ghost him or herself be treated respectfully (its desires broached) and not
> ghosted or abandoned or disappeared again in the act of dealing with the haunt-
> ing, even if the ghost cannot be permitted to take everything over, a compli-
> cated requirement that's especially pertinent with the living who haunt as if
> they were dead. To repeat, for me haunting is not about invisibility or unknow-
> ability per se, it refers us to what's living and breathing in the place hidden from
> view: people, places, histories, knowledge, memories, ways of life, ideas. To
> show what's there in the blind field, to bring it to life on its own terms (and not
> merely to light) is perhaps the radicalization of enlightenments with which I've
> been most engaged. (Gordon 2011, 3)

We have focused on Derwent Whittlesey because he has not been amply
positioned in this institutional history in the slim public record to date.
With the opening of his archives, a fuller accounting was both possible and
necessary. Ever since the duress caused to Whittlesey and his colleagues,
a persistent "something-to-be-done" has haunted US-based geographers
and Harvard-educated students denied the opportunity to learn geography
since 1959.

I thought at that meeting point—in the gracious but careful reckoning with the ghost—we could locate some elements of a practice for moving towards eliminating the conditions that produce the haunting in the first place. For me, this is as much a personal as an intellectual question, and as an intellectual approach it reflects my desire to try to learn how to end the suffering, not merely how to diagnose or diagram or justify or witness. (Gordon 2011, 5)

Some things go willingly into the void; others are thrown in. So it was with the queer history of geography at Harvard, as manifested through the lives of geographers Whittlesey, Kemp, and Ackerman. Queerness was a void. Geography was a void. And Harvard University voided these voids. Yet Harvard could not make itself devoid of these things, as much as it tried, so they kept leaking out—through its secret garden where we met to discuss the archives and secret trials that transpired long before our time on campus.

This story matters not only for Harvard's faculty and students, past, present, and future, and their inability to learn geography, but for the discipline of geography—and particularly geography in the United States. The outcome, secrecy, and political fight surrounding Ackerman's tenure and promotion file continues to haunt geography. The absence of geography at Harvard represents a void within the discipline that seems to tap into geographers' relentless anxieties about their place in the academy. It is worth a collective dwelling in and deepening of our understanding the past where this void takes root with the hope of a different future. We wonder what closures and new openings might be forged through a fuller empirical accounting of and reckoning with the past?

Many people ask what happened to geography at Harvard. Perhaps it is time to ask why geography has not returned, given the discipline's notable growth in popular enrollment, institutionalization, and professionalization in recent decades across the United States and globally. Might it return, as one path to reckoning with this past? Our work is neither done nor can it be completed by us alone. Geography at Harvard remains unfinished business.

NOTES

PROLOGUE

1. Isaiah Bowman, notes on Harvard University's Board of Overseers meeting, October 11, 1948.

2. James Conant, letter to Isaiah Bowman, September 25, 1947.

3. James B. Conant, letter to Paul Buck, January 13, 1948.

4. Alison Mountz first visited Harvard as visiting faculty member on a Mackenzie King fellowship hosted at the Canada Program and Government Department during academic year 2009–2010, returning as William Lyon Mackenzie King Visiting Professor of Canadian Studies in 2015–2016, along with then doctoral candidate in geography at Wilfrid Laurier University, Kira Williams.

5. James Conant, letter to Paul Buck, January 13, 1948.

CHAPTER 1

1. We did discuss with university archivists whether any notes were available that would have been taken at the time Whittlesey's archives were entered. These either did not exist or were not available for review.

2. Harvard University, 1826–1995. Annual Report of the President and Treasurer, 1957, https://guides.library.harvard.edu/harvard-radcliffe-online-historical-ref erence-shelf.

3. Harvard University, 1826–1995. Annual Report of the President and Treasurer, 1927; 1940, https://guides.library.harvard.edu/harvard-radcliffe-online-historical -reference-shelf.

4. Derwent Whittlesey, letter to Paul Buck, June 16, 1943.

CHAPTER 2

1. Emeline McSweeney, letter to Derwent Whittlesey, March 17, 1948.

2. The American Association of Geographers, or AAG, is the leading association of geographers in the United States. It was formerly known as the *Association*

of American Geographers; however, its name changed in 2016 to better reflect and acknowledge its growing internationalism.

3. Derwent Whittlesey, memorandum to Russell Gibson, undated.

4. Clark Currier, letter to Derwent Whittlesey, January 1, 1910.

5. Clark Currier, letter to Derwent Whittlesey, January 1, 1910.

6. Derwent Whittlesey, memorandum to Russell Gibson, undated.

7. Derwent Whittlesey, memorandum to Russell Gibson, undated.

8. Anonymous [Harold Kemp], letter to Derwent Whittlesey, July 20, 1913.

9. Anonymous [Harold Kemp], letter to Derwent Whittlesey, July 20, 1913.

10. Anonymous [Harold Kemp], letter to Derwent Whittlesey, July 20, 1913.

11. Anonymous [Harold Kemp], letter to Derwent Whittlesey, July 20, 1913.

12. Anonymous [Harold Kemp], letter to Derwent Whittlesey, July 20, 1913.

13. Derwent Whittlesey, letter to Allen Wilbur, January 21, 1949.

14. Ruth Whitfield, letter to Derwent Whittlesey, n.d., 1916.

15. Emeline McSweeney, letter to Derwent Whittlesey, August 23, 1916.

16. Emeline McSweeney, letter to Derwent Whittlesey, August 23, 1916.

17. W. W. Atwood, letter to Derwent Whittlesey, n.d., 1917.

18. Ellen Semple, letter to Derwent Whittlesey, undated.

19. Wallace alludes to Derwent's relationship with Harold in a large series of correspondence in the 1920s, even referring to herself in these letters as Derwent's "Maiden Aunt."

20. Derwent Whittlesey, letter to James Fairgrave, December 13, 1922.

21. Derwent Whittlesey, letter to Harlan Barrows, June 14, 1926.

22. William Davis, letter to Derwent Whittlesey, n.d., 1928.

23. [Harold's mother], letter to Derwent Whittlesey, July 2, 1928.

24. [Harold's mother], letter to Derwent Whittlesey, July 2, 1928.

25. Dean Moore, letter to Derwent Whittlesey, April 24, 1928.

26. Derwent Whittlesey, letter to Dean Moore, April 28, 1928.

27. Derwent Whittlesey, letter to Raoul Blanchard, March 1, 1930.

28. Derwent Whittlesey, letter to Olive Whittlesey, September 30, 1931.

29. Derwent Whittlesey, letter to Dean Moore, February 19, 1931.

30. Derwent Whittlesey, letter to Isaiah Bowman, January 7, 1937.

31. Derwent Whittlesey, letter to Isaiah Bowman, January 28, 1937.

32. William Davis, letter to Derwent Whittlesey, June 14, 1932.

33. Derwent Whittlesey, letter to William Christiana, March 19, 1934.

34. Derwent Whittlesey, letter to R. von Kuehnelt-Leddihn, February 18, 1938.

35. Derwent Whittlesey, letter to Paul Buck, June 16, 1943.

36. K. D. Metcalfe, letter to Derwent Whittlesey, January 2, 1940.

37. Max van Rossum Daum, letter to Derwent Whittlesey, October 12, 1947.

38. Derwent Whittlesey, letter to Max van Rossum Daum, October 23, 1947.

39. Derwent Whittlesey resume, April 1948.

40. Derwent Whittlesey, letter to Donald Patton, January 1948.

41. Derwent writes regarding this growth in geography: "Our costs are so high, and our Department has been so small that we had only from one to three [staff] at a time during the years before the war. The G.I. Bill has changed all that, and we now have fifteen." Source: Derwent Whittlesey, letter to Anne Larson, July 29, 1947.

42. Derwent Whittlesey, letter to Henry Bruman, April 12, 1945.

43. Derwent Whittlesey, letter to T. C. Mendenhall, May 7, 1945.

44. Derwent Whittlesey, letter to Charles Colby, April 25, 1947.

45. Derwent Whittlesey, letter to Charles Colby, April 25, 1947.

46. Emeline McSweeney, letter to Derwent Whittlesey, March 17, 1948.

47. Derwent Whittlesey, letter to George Cressey, April 16, 1948.

48. Edward Ackerman, letter to Derwent Whittlesey, March 7, 1948.

49. Harvard University Board of Overseers. Report on Geography. October 11, 1948.

50. Derwent Whittlesey, letter to Grenville Clark, August 12, 1948.

51. Derwent Whittlesey, letter to Dudley Stamp, August 17, 1949.

52. Derwent Whittlesey, letter to Sidman Poole, May 17, 1948.

53. Derwent Whittlesey, letter to L. G. Polspoel, January 3, 1951.

54. Derwent Whittlesey, letter to John Augelli, March 7, 1949.

55. Stephen B. Jones, letter to Derwent Whittlesey, November 18, 1947.

56. Edward Ackerman, letter to Derwent Whittlesey, March 7, 1948.

57. Edward Ackerman, letter to Derwent Whittlesey, February 26, 1948.

58. Derwent Whittlesey, letter to Wellington Jones, April 14, 1952.

59. Dr. Allen Brailey, letter to Derwent Whittlesey, June 13, 1956.

60. Derwent Whittlesey, letter to Robert Steel, November 16, 1956.

61. Slaymaker, Olav, personal correspondence, conversation with Mountz and Williams (April 22, 2022).

CHAPTER 3

1. Emeline McSweeney, letter to Derwent Whittlesey, August 23, 1916.

2. Emeline McSweeney, letter to Derwent Whittlesey, December 6, 1942.

3. Emeline McSweeney, letter to Derwent Whittlesey, February 17, 1948.

4. Kemp taught for a time at Dartmouth College in Hanover, New Hampshire, a couple of hours from Cambridge.

5. Emeline McSweeney, letter to Derwent Whittlesey, September 24, 1916.

6. Some of this personal stationery bore the symbol of the swastika, which—before it fell out of favor for its use by the Nazi party during World War II—was a universal symbol representing peace and good fortune.

7. Emeline McSweeney, letter to Derwent Whittlesey, February 17, 1948.

8. Emeline McSweeney, letter to Derwent Whittlesey, August 23, 1916.

9. Emeline McSweeney, letter to Derwent Whittlesey, August 23, 1916.

10. Emeline McSweeney, letter to Derwent Whittlesey, August 23, 1916.

11. Emeline McSweeney, letter to Derwent Whittlesey, August 23, 1916.

12. Emeline McSweeney, letter to Derwent Whittlesey, August 23, 1916.

13. Emeline McSweeney, letter to Derwent Whittlesey, August 23, 1916.

14. Emeline McSweeney, letter to Derwent Whittlesey, October 22, 1916.

15. Emeline McSweeney, letter to Derwent Whittlesey, February 10, 1945.

16. Emeline McSweeney, letter to Derwent Whittlesey, January 26, 1936. This was a funny revelation for us as professors and instructors of university courses: that blue books have been in use for writing exams on university campuses since at least the 1930s.

17. Emeline McSweeney, letter to Derwent Whittlesey, January 20, 1942.

18. Emeline McSweeney, letter to Derwent Whittlesey, December 6, 1942.

19. Emeline McSweeney, letter to Derwent Whittlesey, December 6, 1942.

20. Emeline McSweeney, letter to Derwent Whittlesey, February 17, 1948

21. Emeline McSweeney, letter to Derwent Whittlesey, February 12, 1939.

22. Emeline McSweeney, letter to Derwent Whittlesey, February 12, 1939.

23. Emeline McSweeney, letter to Derwent Whittlesey, May 19, 1943.

24. Emeline McSweeney, letter to Derwent Whittlesey, January 28, 1944.

25. Emeline McSweeney, letter to Derwent Whittlesey, September 7, 1952.

26. Emeline McSweeney, letter to Derwent Whittlesey, May 19, 1943.

27. Emeline McSweeney, letter to Derwent Whittlesey, January 28, 1944.

28. Emeline McSweeney, letter to Derwent Whittlesey, January 28, 1944.

29. Emeline McSweeney, letter to Derwent Whittlesey, January 28, 1944.

30. Emeline McSweeney, letter to Derwent Whittlesey, July 17, 1936.

31. Emeline McSweeney, letter to Derwent Whittlesey, April 22, 1954.

32. Emeline McSweeney, letter to Derwent Whittlesey, April 22, 1954.

33. Emeline McSweeney, letter to Derwent Whittlesey, January 20, 1942.

34. Emeline McSweeney, letter to Derwent Whittlesey, February 17, 1948.

35. Emeline McSweeney, letter to Derwent Whittlesey, January 29, 1943.

36. Emeline McSweeney, letter to Derwent Whittlesey, "Easter Sunday" (n.d.) April, circa 1935.

37. Emeline McSweeney, letter to Derwent Whittlesey, "Easter Sunday" (n.d.) April, circa 1935.

38. Emeline McSweeney, letter to Derwent Whittlesey, "Easter Sunday" (n.d.) April, circa 1935.

39. Emeline McSweeney, letter to Derwent Whittlesey, "Easter Sunday" (n.d.) April, circa 1935.

40. Emeline McSweeney, letter to Derwent Whittlesey, "Easter Sunday" (n.d.) April, circa 1935.

41. Emeline McSweeney, letter to Derwent Whittlesey, "Easter Sunday" (n.d.) April, circa 1935.

42. Emeline McSweeney, letter to Derwent Whittlesey, "Easter Sunday" (n.d.) April, circa 1935.

43. Emeline McSweeney, letter to Derwent Whittlesey, "Easter Sunday" (n.d.) April, circa 1935.

44. Emeline McSweeney, letter to Derwent Whittlesey, "Easter Sunday" (n.d.) April, circa 1935.

45. Emeline McSweeney, letter to Derwent Whittlesey, June 17, 1948.

46. Emeline McSweeney, letter to Derwent Whittlesey, "Easter Sunday" (n.d.) April, circa 1935.

47. Emeline McSweeney, letter to Derwent Whittlesey, "Easter Sunday" (n.d.) April, circa 1935.

48. Emeline McSweeney, letter to Derwent Whittlesey, June 17, 1948.

49. Emeline McSweeney, letter to Derwent Whittlesey, June 17, 1948.

50. Emeline McSweeney, letter to Derwent Whittlesey, "Easter Sunday" (n.d.) April, circa 1935.

51. Emeline McSweeney, letter to Derwent Whittlesey, "Easter Sunday" (n.d.) April, circa 1935.

52. Emeline McSweeney, letter to Derwent Whittlesey, April 22, 1954.

53. Emeline McSweeney, letter to Derwent Whittlesey, April 22, 1954.

54. Emeline McSweeney, letter to Derwent Whittlesey, January 29, 1943.

55. Emeline McSweeney, letter to Derwent Whittlesey, January 29, 1943.

56. Emeline McSweeney, letter to Derwent Whittlesey, January 29, 1943.

57. Emeline McSweeney, letter to Derwent Whittlesey, January 29, 1943.

58. Emeline McSweeney, letter to Derwent Whittlesey, January 28, 1944.

59. Emeline McSweeney, letter to Derwent Whittlesey, January 28, 1944.

60. Emeline McSweeney, letter to Derwent Whittlesey, January 28, 1944.

61. Emeline McSweeney, letter to Derwent Whittlesey, January 28, 1944.

62. Emeline McSweeney, letter to Derwent Whittlesey, April 22, 1954.

63. Emeline McSweeney, letter to Derwent Whittlesey, n.d., September 1956.

64. Emeline McSweeney, letter to Derwent Whittlesey, n.d., September 1956.

65. Emeline McSweeney, letter to Derwent Whittlesey, n.d., September 1956.

66. Emeline McSweeney, letter to Derwent Whittlesey, n.d., September 1956.

67. Emeline McSweeney, letter to Derwent Whittlesey, February 10, 1945.

68. Derwent Whittlesey, letter to Mrs. Robert Sickels, August 11, 1955. We recognize the ableism of this language, seeking to situate ableist language in their time, alongside our knowledge and ability to critique it in ours.

CHAPTER 4

1. James B. Conant, letter to Paul Buck, January 13, 1948.

2. Paul Buck, letter to Kirk Bryan, February 9, 1948.

3. Clifford Moore, letter to Derwent Whittlesey, April 24, 1928.

4. Derwent Whittlesey, letter to Clifford Moore, April 20, 1928.

5. Clifford Moore, letter to Derwent Whittlesey, April 24, 1928.

6. Derwent Whittlesey, letter to Raoul Blanchard, March 1, 1930. We assume that the university administration was not aware of the long-term romantic relationship between Whittlesey and Kemp at this time; we cannot know if any colleagues in the department would have been aware.

7. Clifford Moore, letter to Derwent Whittlesey, February 21, 1931.

8. Clifford Moore, letter to Derwent Whittlesey, April 24, 1928; Derwent Whittlesey, letter to William Christina, March 19, 1934.

9. Agreement between Harvard University and Alexander Hamilton Rice on creation of Institute for Geographical Exploration, June 18, 1930. Harvard University Archives.

10. Agreement between Harvard University and Alexander Hamilton Rice on creation of Institute for Geographical Exploration, June 18, 1930. Harvard University Archives.

11. Bowman was one of the Institute's most ardent opponents, explained in a secret letter he instructed be opened after Hamilton's death (1956). Details can be found in Isaiah Bowman, secret letter on Alexander Hamilton Rice, July 27, 1937.

12. James Conant, letter to Derwent Whittlesey, November 13, 1934; Donald McLaughlin, letter to James Conant, January 16, 1935.

13. Donald McLaughlin, letter to James B. Conant, January 16, 1935.

14. Derwent Whittlesey, letter to Howard Meyerhoff, May 9, 1938.

15. Donald McLaughlin, letter to George Birkoff, April 13, 1939.

16. Derwent Whittlesey, letter to Henry Bruman, April 12, 1945; Derwent Whittlesey, letter to Anne Larson, July 29, 1947.

17. Derwent Whittlesey, letter to Paul Buck, March 27, 1947.

18. Harold Kemp, letter to Local Board #47 Phillips Brooks House, December 11, 1940.

19. Derwent Whittlesey, letter to Matthew McClure, January 3, 1944.

20. One example of this conflict was the rejection of the appointment of Edward Ackerman as an assistant professor in 1939. Conant rejected the promotion not due to Ackerman's qualifications but because geology would lose a new position if he were appointed. For more information, see George Chase, letter to Derwent Whittlesey, March 14, 1939.

21. Paul Herman Buck, letter to Derwent Whittlesey, May 15, 1945.

22. David Bailey, letter to Derwent Whittlesey, May 5, 1947.

23. Derwent Whittlesey, letter to Eric Archdeacon, March 28, 1947.

24. Derwent Whittlesey, letter to Richard Hartshorne, June 3, 1947.

25. James Conant, letter to Roderick Peattie, February 14, 1935.

26. George Chase, personal meeting with Derwent Whittlesey, March 16, 1939.

27. Derwent Whittlesey, letter to Max van Rossum Daum, October 12, 1947.

28. Derwent Whittlesey, letter to Max van Rossum Daum, October 23, 1947.

29. Derwent Whittlesey, letter to Eric Archdeacon, March 28, 1947.

30. Derwent Whittlesey, letter to Hubert Schenck, August 1, 1947.

31. See Isaiah Bowman, letter to James Conant, 10 October 1947; and James Conant, letter to Isaiah Bowman, December 4, 1947.

32. James Conant, letter to Paul Herman Buck, January 13, 1948.

33. Derwent Whittlesey, letter to Montgomery Bradley, January 11, 1951.

34. Derwent Whittlesey, letter to Charles Leonard, September 27, 1951.

35. Derwent Whittlesey, letter to Richard Logan, December 12, 1951.

36. Derwent Whittlesey, letter to Thomas McAvoy, March 20, 1953.

37. Nathaniel Pusey, letter to Derwent Whittlesey, April 20, 1954.

38. Derwent Whittlesey, letter to Samuel Emory, January 4, 1954.

39. Derwent Whittlesey, letter to McGeorge Bundy, July 1, 1954.

40. Derwent Whittlesey, letter to William Talbot, February 20, 1956.

41. Derwent Whittlesey, letter to Charles Bird, February 15, 1952.

42. Derwent Whittlesey, letter to Robert Johnson, November 24, 1948.

43. Marland Billings, letter to Paul Herman Buck, January 21, 1950.

44. Emeline McSweeney, letter to Derwent Whittlesey, February 10, 1945.

45. Emeline McSweeney, letter to Derwent Whittlesey, March 17, 1948.

46. Emeline McSweeney, letter to Derwent Whittlesey, March 17, 1948.

47. Derwent Whittlesey, letter to Charles Colby, April 25, 1947.

48. Emeline McSweeney, letter to Derwent Whittlesey, June 17, 1948.

49. Emeline McSweeney, letter to Derwent Whittlesey, June 17, 1948.

50. See Derwent Whittlesey, letter to Samuel Emory, January 4, 1954; and Derwent Whittlesey, letter to a prospective student, March 8, 1955.

CHAPTER 5

1. In this endeavor, we would like to thank Johns Hopkins University archivist Jim Limpert, who assisted us during our research. Having worked with Isaiah Bowman's records since 1984, Jim actually met and also assisted Neil Smith during his studies there from at least 1984 to 1991. Jim shared stories of not only Neil with us but also Isaiah, who he claimed was a peculiar and strangely egotistical figure. Jim also informed us that it was likely Neil who had the access requirements to his archives overturned in the 1980s, thereby paving the way for our own subsequent access.

2. *National Cyclopedia of American Biography*, s.v. "Isaiah Bowman." James T. White & Co. Publishers (1950).

3. Isaiah Bowman, letter to James Love, July 5, 1939.

4. Isaiah Bowman, letter to James Love, July 5, 1939.

5. Isaiah Bowman, letter to James Love, July 5, 1939; *National Cyclopedia of American Biography*, s.v. "Isaiah Bowman." James T. White & Co. Publishers (1950).

6. Isaiah Bowman, letter to James Love, July 5, 1939; *National Cyclopedia of American Biography*, s.v. "Isaiah Bowman." James T. White & Co. Publishers (1950).

7. Isaiah Bowman, letter to James Love, July 5, 1939; *National Cyclopedia of American Biography*, s.v. "Isaiah Bowman." James T. White & Co. Publishers (1950).

8. Isaiah Bowman, letter to James Love, July 5, 1939; *National Cyclopedia of American Biography*, s.v. "Isaiah Bowman." James T. White & Co. Publishers (1950).

9. Isaiah Bowman, letter to James Love, July 5, 1939; *National Cyclopedia of American Biography*, s.v. "Isaiah Bowman." James T. White & Co. Publishers (1950).

10. *National Cyclopedia of American Biography*, s.v. "Isaiah Bowman." James T. White & Co. Publishers (1950).

11. *National Cyclopedia of American Biography*, s.v. "Isaiah Bowman." James T. White & Co. Publishers (1950).

12. *National Cyclopedia of American Biography*, s.v. "Isaiah Bowman." James T. White & Co. Publishers (1950).

13. *National Cyclopedia of American Biography*, s.v. "Isaiah Bowman." James T. White & Co. Publishers (1950).

14. Isaiah Bowman, Harvard University Board of Overseers Meeting Notes, October 11, 1948.

15. Isaiah Bowman, letter to James Love, July 5, 1939.

16. *National Cyclopedia of American Biography*, s.v. "Isaiah Bowman." James T. White & Co. Publishers (1950).

17. James Conant, letter to Isaiah Bowman, January 20, 1936.

18. James Conant, letter to Isaiah Bowman, September 25, 1947.

19. Isaiah Bowman, letter to James Conant, November 26, 1947, 1–2.

20. Isaiah Bowman, letter to James Conant, November 26, 1947, 2.

21. James Conant, letter to Isaiah Bowman, December 4, 1947.

22. Isaiah Bowman, letter to James Conant, December 8, 1947, 1–2.

23. Isaiah Bowman, letter to James Conant, December 8, 1947, 2.

24. Isaiah Bowman, letter to James Conant, December 9, 1947.

25. Isaiah Bowman, letter to James Conant, December 9, 1947, 2.

26. Isaiah Bowman, letter to James Conant, September 24, 1948.

27. Isaiah Bowman, letter to James Conant, September 24, 1948; Isaiah Bowman, note on the Brookhaven Laboratory Conference, October 13, 1948.

28. Isaiah Bowman, letter to James Conant, September 24, 1948; Isaiah Bowman, note on the Brookhaven Laboratory Conference, October 13, 1948.

29. Isaiah Bowman, letter to James Conant, September 24, 1948; Isaiah Bowman, note on the Brookhaven Laboratory Conference, October 13, 1948.

30. Isaiah Bowman, notes on Harvard University's Board of Overseers meeting, October 11, 1948.

31. Isaiah Bowman, notes on Harvard University's Board of Overseers meeting, October 11, 1948, 3–4.

32. Isaiah Bowman, notes on Harvard University's Board of Overseers meeting, October 11, 1948, 4.

33. Isaiah Bowman, notes on Harvard University's Board of Overseers meeting, October 11, 1948, 5.

34. Isaiah Bowman, notes on Harvard University's Board of Overseers meeting, October 11, 1948, 5–6.

35. Isaiah Bowman, notes on Harvard University's Board of Overseers meeting, October 11, 1948, 6.

36. Isaiah Bowman, letter to Derwent Whittlesey, March 2, 1949.

37. Derwent Whittlesey, letter to Isaiah Bowman, January 7, 1937.

38. Derwent Whittlesey, letter to Isaiah Bowman, April 29, 1937.

39. Derwent Whittlesey, letter to Isaiah Bowman, January 4, 1941.

40. Isaiah Bowman, letter to Derwent Whittlesey, December 4, 1945.

41. Derwent Whittlesey, letter to Isaiah Bowman, March 24, 1944.

42. Isaiah Bowman, letter to Derwent Whittlesey, March 28, 1944.

43. Derwent Whittlesey, letter to Isaiah Bowman, October 7, 1949.

44. Isaiah Bowman, letter to Derwent Whittlesey, October 13, 1949.

45. Isaiah Bowman, letter to Paul Buck, May 12, 1948.

46. Edward Ullman, letter to Isaiah Bowman, February 25, 1949.

47. Isaiah Bowman, letter to Edward Ullman, March 2, 1949.

48. Isaiah Bowman, letter to Edward Ullman, March 10, 1949.

49. Isaiah Bowman, letter to Edward Ullman, May 27, 1949.

50. Isaiah Bowman, letter to Edward Ullman, May 27, 1949.

CHAPTER 6

1. Edward Ackerman, Letter, March 7, 1948. He also notes that Peter Roll, "one of Harold's former students," is organizing alumni "behind the scenes" to fight for geography.

2. Senator Joe McCarthy, letter to US President, February 3, 1953, p. 2.

3. Senator Joe McCarthy, letter to US President, February 3, 1953, p2.

4. Conant's sexual similes—such as likening Marxism to pornography—lend insight into his person.

5. Notice the use of "talented students" in his writing, not "poor." His goal is not to eliminate or reduce poverty but "cull" aristocracy.

6. At the same time, Conant did create new programs at Harvard during his administration, which contradicted this assertion (i.e., public policy, history/philosophy of science, education). Perhaps he saw these new programs as necessary and, therefore, acceptable for expansion.

7. Bowman's appeal, in the letter, to note geography was *not* easy to Conant may reveal his astuteness in working with Conant in this regard.

8. Terris Moore, letter to President Lawrence Lowell, October 27, 1927.

9. Isaiah Bowman, letter to James Conant, September 6, 1941.

10. Isaiah Bowman, letter to James Conant, September 6, 1941.

11. Isaiah Bowman, archival memo titled "BROOKHAVEN LABORATORY CONFERENCE," October 13, 1948.

12. Isaiah Bowman, letter to James Conant, September 14, 1936.

13. James Conant, letter to Isaiah Bowman, January 22, 1945.

14. James Conant, letter to Isaiah Bowman, January 22, 1945.

15. Isaiah Bowman, letter to James Conant, January 26, 1945; James Conant, letter to Isaiah Bowman, February 2, 1945.

16. James Conant, letter to Isaiah Bowman, September 25, 1947.

17. Derwent Whittlesey, letter to Willard Miller, April 16, 1948.

18. Derwent Whittlesey, letter to Grenville Clark, August 17, 1948.

19. Derwent Whittlesey, letter to Robert Johnson, November 24, 1948.

20. Paul Herman Buck, letter to Marland Billings, August 12, 1948; Derwent Whittlesey, letter to Marland Billings, August 20, 1948.

21. Bowman, who was elected to the Board of Overseers in 1948, took detailed notes of the report and its related October 11 meeting, which he left in his personal archives. For more information, see Isaiah Bowman, notes from Overseers Meeting, October 11, 1948.

22. Derwent Whittlesey, letter to Montgomery Bradley, January 11, 1951.

23. Derwent Whittlesey, letter to Dudley Stamp, April 17, 1952.

24. Meeting of the Administrative Committee of Harvard University, April 20, 1953.

25. Nathaniel Pusey, letter to Charles Bird, April 20, 1954.

26. James Conant, letter to Paul Herman Buck, January 13, 1948

27. James Conant, letter to Alexander Hamilton Rice, September 15, 1951.

28. David Bailey, letter to James Conant, October 5, 1951.

29. Alexander Hamilton Rice, letter to James Conant, September 24, 1951, James Conant, letter to Terris Moore, January 10, 1952.

30. James Conant, letter to Paul Herman Buck, October 23, 1952.

31. Derwent Whittlesey, letter to Dudley Stamp, July 28, 1952.

32. James Conant, letter to Dudley Stamp, January 23, 1933.

33. Nathaniel Pusey, letter to Charles Bird, April 20, 1954.

34. Charles Brooks, letter to Lewis Don Leet, February 18, 1955.

CHAPTER 8

1. Edward Ackerman to Derwent S. Whittlesey, n.d.).

2. Derwent S. Whittlesey to Edward Ackerman, September 30, 1948.

3. Derwent S. Whittlesey to Edward Ackerman, September 30, 1948.

4. Edward Ackerman to Derwent S. Whittlesey, September 12, 1948.

5. Edward Ackerman to Derwent S. Whittlesey, September 12, 1948.

6. Edward Ackerman to Derwent S. Whittlesey, September 12, 1948.

7. Derwent Whittlesey to Edward Ackerman, March 6, 1949.

8. The Geography Building now houses the Department of East Asian Languages and Civilizations and the Harvard-Yenching Library (Harvard Alumni Bulletin October 13, 1951; Harvard University 1954).

9. Letter from Edward Ackerman to Dean Paul Buck, August 1, 1948. Ackerman archives, American Heritage Center, University of Wyoming.

10. Derwent Whittlesey, letter to Charles Leonard, September 27, 1951.

11. Letter from Derwent Whittlesey to Edward Ackerman, August 2, 1945. Archives of Edward Ackerman

12. Derwent Whittlesey, letter to Hubert Schenck, August 1, 1947.

13. Isaiah Bowman, letter to Kirk Bryan, March 22, 1948.

REFERENCES

Ackerman, Edward A. 1957. "Derwent Stainthorpe Whittlesey." *Geographical Review* 47, no. 3: 443–445.

Agnew, John. 1994. "The Territorial Trap: The Geographical Assumptions of International Relations Theory." *Review of International Political Economy* 1, no. 1: 53–80.

Ahmed, Sara. 2021. *Complaint!* Durham: Duke University Press.

Anderson, Elizabeth. 2015. "Feminist Epistemology and Philosophy of Science." *Stanford Encyclopedia of Philosophy*.

Anderson, Isaac. 1944. "Affairs of Destiny by Georges Simenon." *New York Times*, November 26, 1944. Accessed July 18, 2023. https://timesmachine.nytimes.com /timesmachine/1944/11/26/87477614.pdf?pdf_redirect=true&ip=0.

Ashworth, Lucian. 2020. "Chronicle of a Death Foretold? The 1953–4 CFR Study Group Meeting and the Decline of International Thought." *The International History Review* 42, no. 3: 656–671.

Ashworth, Lucian. 2021. "A Forgotten Environmental International Relations: Derwent Whittlesey's International Thought." *Global Studies Quarterly* 1, no. 2: 1–10.

Associated Press. 1978. "James B. Conant Is Dead at 84; Harvard President for 20 Years; From Chemistry to Top Post." *New York Times*, February 12, 1978.

Barlett, Paul D. 1983. *James Bryant Conant, 1893–1978. A Biographical Memoir*. Washington, DC: National Academy of Sciences.

Barnes, Trevor J. 2001. "Lives Lived and Lives Told: Biographies of Geography's Quantitative Revolution." *Environment and Planning D: Society and Space* 19, no. 4: 409–429.

Barnes, Trevor J. 2006. "Geographical Intelligence: American Geographers and Research and Analysis in the Office of Strategic Services 1941–1945." *Journal of Historical Geography* 32, no. 1: 149–168.

Barnes, Trevor J. 2016. "American Geographers and World War II: Spies, Teachers, and Occupiers." *Annals of the Association of American Geographers* 106, no. 3: 543–550.

Barnes, Trevor J., and Matthew Farish. 2006. "Between Regions: Science, Militarism, and American Geography from World War to Cold War." *Annals of the Association of American Geographers* 96, no. 4: 807–826.

Benavente, Gabby, and Julian Gill-Peterson. 2019. "The Promise of Trans Critique: Susan Stryker's Queer Theory." *A Journal of Gay and Lesbian Studies* 25, no. 1: 23–28.

Biddle, Justin. 2011. "Putting Pragmatism to Work in the Cold War: Science, Technology, and Politics in the Writings of James B. Conant." *Studies in History and Philosophy of Science Part A* 42, no. 4: 552–561.

Board of Overseers, Harvard University. 1948. *Report on Geography at Harvard University*, October 11. Cambridge, MA: Harvard University.

Bordo, Susan. 1987. *The Flight to Objectivity: Essays on Cartesianism and Culture*. Albany: SUNY Press.

Bowman, Isaiah. Collection: Isaiah Bowman papers. Johns Hopkins University Archives, Baltimore, MD, USA. https://aspace.library.jhu.edu/repositories/3/resources/69.

Bowman, Isaiah. 1937. Letter to no one, released after death. July 27, 1937. Johns Hopkins University Archives, Baltimore, MD, USA.

Bowman, Isaiah. 1948. Notes from Harvard University Board of Overseers Meeting, October 11, 1948. John Hopkins University Archives, Baltimore, MD, USA.

Braukman, Stacy. 2012. *Communists and Perverts under the Palms: The Johns Committee in Florida, 1956–1965*. Gainesville: University Press of Florida.

Brown, Michael P. 2005. *Closet Space: Geographies of Metaphor from the Body to the Globe*. New York: Routledge.

Browning, Robert. 1895. "VI Reading a Book, Under the Cliff." In *The Complete Poetic and Dramatic Works of Robert Browning: Cambridge Edition*, edited by Horace E. Scudder, 374–375. Houghton: Mifflin and Company.

Butler, Judith. 1990. *Gender Trouble*. New York: Routledge.

Center for Geographical Analysis. 2017. "About: Center for Geographical Analysis." Center for Geographical Analysis, Harvard University. Accessed October 26, 2017. https://gis.harvard.edu/about.

Cho, Isabella. 2021. "'We'll Keep Telling the Truth': A Century Later, Harvard Affiliates Continue Pushing Harvard to Address 1920 Secret Court." *The Crimson*, May 27, 2021. Accessed 18 July 2023. https://www.thecrimson.com/article/2021/5/27/century-after-secret-court/.

Code, Lorraine. 1991. *What Can She Know? Feminist Theory and the Construction of Knowledge*. Ithaca, NY: Cornell University Press.

Cohen, Saul B. 1988. "Reflections on the Elimination of Geography at Harvard, 1947–1951." *Annals of the Association of American Geographers* 78, no. 1: 148–151.

Cohen, Saul B. 2003. *Geopolitics of the World System*. New York: Rowman & Littlefield Publishers.

"College Dooms Major in Geographical Field." *Harvard Crimson*, March 4, 1948. Accessed July 18, 2023. https://www.thecrimson.com/article/1948/3/4/college -dooms-major-in-geographical-field/.

Conant, James B. Papers of James Bryant Conant, 1862–1987. 2020. Harvard University Archives, Cambridge, MA, USA. https://hollisarchives.lib.harvard.edu /repositories/4/resources/4091.

Conant, James B. 1948. Letter to Provost of Faculty of Arts and Sciences Dean Paul Buck. Harvard University, January 13, 1948. Harvard University Archives, Cambridge, MA, USA.

Conant, James B. 1948. *Education in a Divided World*. Cambridge, MA: Harvard University Press.

Conant, James B. 1953. *Education and Liberty: The Role of Schools in Modern Democracy*. Cambridge, MA: Harvard University Press.

Conant, Jennet. 2017. *Man of the Hour: Warrior Scientist*. New York: Simon and Schuster.

Conant, James Bryant. 1948. *Education in a Divided World: The Function of the Public Schools in Our Unique Society*. Cambridge: Harvard University Press.

Davis, William Morris. 1924. "The Progress of Geography in the United States." *Annals of the Association of American Geographers* 14, no. 4: 159–215.

Department of Geology and Geography, Harvard University. 1928. *Memorandum Concerning Plan for Expansion of Work in Geology and Mining*, November 27, 1928.

Derrida, Jacques. 2012. *Specters of Marx: The State of the Debt, the Work of Mourning and the New International*. New York: Routledge.

Diamond, Sigmund. 1992. *Compromised Campus: The Collaboration of Universities with the Intelligence Community, 1945–1955*. Oxford: Oxford University Press.

Division of Geological Sciences, Harvard University. 1944. "Post-War Plans for Geology: Report to the Dean of the Faculty of Arts and Sciences." July 1, 1944.

Domosh, Mona. 1991. "Toward a Feminist Historiography of Geography." *Transactions of the Institute of British Geographers* 16, no. 1: 9104.

Fricker, Miranda. 2007. *Epistemic Injustice: Power and the Ethics of Knowing*. Oxford University Press.

Gieseking, Jen Jack. 2015. "Useful In/stability: The Dialectical Production of the Social and Spatial Lesbian Herstory Archives." *Radical History Review* 2015, no. 122: 25–37.

Gladys, Wrigley. 1951. *Isaiah Bowman*. New York: American Geographical Society.

Glick, Thomas. 1988. "Before the Revolution: Edward Ullman and the Crisis of Geography at Harvard, 1949–1950." In *Geography in New England*, edited by J. Harmon and T. Rickard, 49–62. New Britain: New England St. Lawrence Valley Geographical Society.

Gordon, Avery. 2008. *Ghostly Matters: Haunting and the Sociological Imagination*. Minneapolis: University of Minnesota Press.

Gordon, Avery. 2011. "Some Thoughts on Haunting and Futurity." *Borderlands* 10, no. 2: 1–21.

Graves, Karen. 2009. *And They Were Wonderful Teachers: Florida's Purge of Gay and Lesbian Teachers*. University of Illinois Press.

Haraway, Donna. 1991. "Situated Knowledges: The Science Question in Feminism and the Privilege of Partial Perspective." In *Simians, Cyborgs, and Women*, 183–202. New York: Routledge.

Harvard Alumni Bulletin. 1951. "Support Withdrawn." October 13, 1951.

Harvard University. 1826–1995. *Annual Reports of the President and Treasurer*. https:// guides.library.harvard.edu/harvard-radcliffe-online-historical-reference-shelf.

Harvard University. 1928. "Reports of the President and the Treasurer of Harvard College 1927–1928." Accessed October 26, 2017. http://nrs.harvard.edu/urn -3:hul.arch:15004.

Harvard University. 1940. "Issue Containing the Report of the President of Harvard College and Reports of Departments for 1938–1939."

Harvard University. 1941. "Reports of the President and the Treasurer of Harvard College 1940–1941." Accessed October 26, 2017. http://nrs.harvard.edu/urn -3:hul.arch:15003.

Harvard University. 1942. "Reports of the President and the Treasurer of Harvard College 1941–1942." Accessed October 26, 2017. http://nrs.harvard.edu/urn -3:hul.arch:15003.

Harvard University. 1943. "Reports of the President and the Treasurer of Harvard College 1942–1943." Accessed October 26, 2017. http://nrs.harvard.edu/urn -3:hul.arch:15003.

Harvard University. 1946. "Reports of the President and the Treasurer of Harvard College 1945–1946." Accessed October 26, 2017. http://nrs.harvard.edu/urn -3:hul.arch:15003.

Harvard University. 1947. "Reports of the President and the Treasurer of Harvard College 1946–1947." Accessed October 26, 2017. http://nrs.harvard.edu/urn -3:hul.arch:15003.

Harvard University. 1948. "Reports of the President and the Treasurer of Harvard College 1947–1948." Accessed October 26, 2017. http://nrs.harvard.edu/urn -3:hul.arch:15003.

Harvard University. 1954. *The Behavioral Sciences at Harvard.* Cambridge: Harvard University Press.

Harvard University. 1961. "Reports of the President and the Treasurer of Harvard College 1960–1961." Accessed October 26, 2017. http://nrs.harvard.edu/urn -3:hul.arch:15008.

Harvard University. 2023. "Louis Agassiz." Accessed July 18, 2023. https://eps .harvard.edu/louis-agassiz.

Haslanger, Sally. 2000. "Gender and Race: (What) Are They? (What) Do We Want Them to Be?" *Noûs* 34, no. 1: 31–55.

Hayden, Dolores. 1997. *The Power of Place: Urban Landscapes as Public History.* Cambridge, MA: MIT Press.

Hershberg, James G. 1993. *James B. Conant: Harvard to Hiroshima and the Making of the Nuclear Age.* New York: Alfred A. Knopf.

Hershberg, James G. 2019. "Man of the Hour: James B. Conant, Warrior Scientist." *Journal of Cold War Studies* 21, no. 1: 186–189.

Hookway, Christopher. 2010. "Some Varieties of Epistemic Injustice: Reflections on Fricker." *Episteme* 7, no. 2: 151–163.

Johnson, David. 2006. *The Lavender Scare: The Cold War Persecution of Gays and Lesbians in the Federal Government.* Chicago: University Chicago Press.

Kappa Kappa Gamma. 1954. "Four Receive New Educational Awards." *The Key,* October. Accessed July 18, 2023. https://wiki.kkg.org/images/0/08/THE_KEY _VOL_71_NO_3_OCT_1954.pdf.

Kappa Kappa Gamma. 1959. "In Memoriam." *The Key,* Mid-Winter. Accessed July 18, 2023. https://wiki.kkg.org/images/9/93/THE_KEY_VOL_76_NO_1 _MID-WINTER_1959.pdf.

Kemp, Harold. 1960. *Last Will and Testament of Harold Kemp.* Cambridge: Middlesex County Probate and Family Court, September 15.

Kevles, Daniel J. 1977. "The National Science Foundation and the Debate over Postwar Research Policy, 1942–1945. A Political Interpretation of Science-the Endless Frontier." *Isis* 68, no. 241: 5–26.

Kingsbury, Paul, and Anna Secor, eds. 2021. *A Place More Void.* Lincoln: University of Nebraska Press.

Larsen, S., and J. Johnson. 2012. "In Between Worlds: Place, Experience, and Research in Indigenous Geography." *Journal of Cultural Geography* 29: 1–13.

Lookabaugh, Lara. 2022. "Body of Evidence: Time and Desire in Embodied Archives." *Qualitative Inquiry* 28, no. 10: 1039–1050.

MacDonald, Glen. 2016. "The End(s) of Geography?" *Newsletter of the American Association of Geographers*, July 1, 2016. Accessed September 22. 2019. https://www.aag.org/the-ends-of-geography.

Marston, Sallie A. 2000. "The Social Construction of Scale." *Progress in Human Geography* 24, no. 2: 219–242.

Martin, Geoffrey. 1980. *The Life and Thought of Isaiah Bowman*. New York: Archon Books.

Martin, Geoffrey. 1988. "On Whittlesey, Bowman and Harvard." *Annals of the Association of American Geographers* 78, no. 1: 152–158.

McKittrick, Katherine, and Clyde Woods. 2007. *Black Geographies and the Politics of Place*. Toronto: Between the Lines.

Mitchell, Don. 2014. "Neil Smith, 1954–2012: Marxist Geographer." *Annals of the Association of American Geographers* 104, no. 1: 215–222.

Morris, Rita. 1962. "An Examination of Some Factors Related to the Rise and Decline of Geography as a Field of Study at Harvard, 1638–1948." Doctoral diss., Harvard University.

Mountz, Alison, and Kira Williams. 2021. "Derwent's Ghost: The Haunting Silences of Geography's History at Harvard." In *A Place More Void*, edited by Paul Kingsbury and Anna Secor, 132–149. Lincoln, NE: University of Nebraska Press.

Mountz, Alison, and Kira Williams. 2023. "Let Geography Die: The Rise, Fall, and 'Unfinished Business' of Geography at Harvard." *Annals of the Association of American Geographers* 113, no. 8: 1977–2002.

Nelson, Lynn. 1992. *Who Knows: From Quine to a Feminist Empiricism*. Philadelphia: Temple University Press.

"New Committee to Sift Geography Meets Today: Buck Names Group to Reopen Year-Old Problem of Field; Mather and Ullman on List." *Harvard Crimson*, May 11, 1949. Accessed July 18, 2023. https://www.thecrimson.com/article/1949/5/11/new-committee-to-sift-geography-meets/.

Noxolo, Patricia. 2022. "Geographies of Race and Ethnicity 1: Black Geographies." *Progress in Human Geography* 46, no. 5: 1232–1240.

O'Connell, Jeffrey, and Thomas E. O'Connell. 1997. "James B. Conant: A Giant on Academe's Left, Right and Center, Book Review of James B. Conant, Harvard to Hiroshima and the Making of the Nuclear Age by James G. Hershberg." *Brigham Young University Education and Law Journal* 1997, no. 1: 109–123.

Oswin, Natalie. 2020. "An Other Geography." *Dialogues in Human Geography* 10, no. 1: 9–18.

O'Tuathail Gearóid. 1996. *Critical Geopolitics*. Minneapolis: University of Minnesota Press.

Painter, Joe. 1995. *Politics, Geography and 'Political Geography': A Critical Perspective*. London Edward Arnold.

Patton, Donald J. 1974. "Obituary: Edward A. Ackerman (1911–1973)." *Geographical Review* 64, no. 1: 150–153.

Pettis, Ruth M. 2004. "Uranianism." *GLBTQ Archive*. Accessed July 18, 2023. http://www.glbtqarchive.com/ssh/Uranianism_S.pdf.

Roll, Peter. "Elimination of Geography Meets Disapproval." *Harvard Crimson* (Cambridge, MA), March 9, 1948. Accessed July 18, 2023. https://www.thecrimson.com/article/1948/3/9/elimination-of-geography-meets-disapproval-to/

Rose, Gillian. 1993. *Feminism and Geography: The Limits of Geographical Knowledge*. London: Polity.

Schnur, James A. 1997. "Closet Crusaders: The Johns Committee and Homophobia." In *Carryin' on in the Lesbian and Gay South*, edited by John Howard, 109–123. New York: New York University Press.

Serano, Julia. 2016. *Whipping Girl: A Transsexual Woman on Sexism and the Scapegoating of Femininity*. London: Hachette UK.

Shinkle, Peter. 2022. *Ike's Mystery Man: The Secret Lives of Robert Cutler*. New York: Steerforth Press/Truth to Power.

Slayermaker, Olav. "Interview of Olav Slayermaker," by Leonora King. Public Domain. February 7, 2020. Accessed July 18, 2023.

Smith, Neil. 1987. "Academic War over the Field of Geography: The Elimination of Geography at Harvard, 1947–1951." *Annals of the Association of American Geographers* 77, no. 2: 155–172.

Smith, Neil. 2003. *American Empire: Roosevelt's Geographer and the Prelude to Globalization*. Berkeley: University of California Press.

Stamp, Dudley. 1952. *Report on the Status of Geography at Harvard University*. October 12. Cambridge, MA: Harvard University.

Syrett, Nicholas. 2021. *An Open Secret: The Family Story of Robert and John Gregg Allerton*. Chicago: University of Chicago Press.

Talkoff, Emma. "All Over the Map." *Harvard Crimson* (Cambridge, MA), April 21, 2016. Accessed July 18, 2023. https://www.thecrimson.com/article/2016/4/21/map-collections/

Tolya-Kelly, Divya. 2006. "Affect: An Ethnocentric Encounter? Exploring the 'Universalist' Imperatives of Emotional/Affectual Geographies." *Area* 38, no. 2: 213–217.

Townsend, Kim. 1996. *Manhood at Harvard: Williams James and Others*. New York: W.W. Norton & Company.

"United States Census, 1900." Database with images. FamilySearch. http:// FamilySearch.org : June 18, 2024. Citing NARA microfilm publication T623. Washington, D.C.: National Archives and Records Administration, n.d.

White, Gilbert F. 1974. "Edward A. Ackerman, 1911–1973." *Annals of the Association of American Geographers* 62, no. 2: 297–309.

Whittlesey, Derwent. Papers of Derwent Stainthorpe Whittlesey, 1908–1956. Harvard University Archives, Cambridge, MA, USA.

Whittlesey, Derwent. 1935. "Political Geography." *Education* 55: 293–298.

Whittlesey, Derwent. 1939. *The Earth and the State*. New York: Henry Holt and Company.

Whittlesey, Derwent. 1956. *Last Will and Testament of Derwent Whittlesey*. Cambridge: Middlesex County Probate and Family Court, December 27.

Wilder, Craig S. 2013. *Ebony and Ivory: Race, Slavery, and the Troubled History of America's Universities*. New York: Bloomsbury Press

Wilson, Matthew W. 2017. *New Lines: Location-Aware Futures and the Map*. Minneapolis: University of Minnesota Press.

Wright, Richard, and Natalie Koch. 2009. "Ivy League and Geography in the US." In *International Encyclopedia of Human Geography*, edited by Rob Kitchin and Nigel Thrift, 616–621. Amsterdam: Elsevier.

Wright, William. 2006. *Harvard's Secret Court: The Savage 1920 Purge of Campus Homosexuals*. New York: St. Martin's Griffin.

Young, Iris. 1990. *Throwing Like a Girl and Other Essays in Feminist Political Theory*. Bloomington: Indiana University Press.

Page numbers followed by "f" denote a figure, "t" a table, and "n" a note, respectively.

Publisher contact:
The MIT Press
Massachusetts Institute of Technology
77 Massachusetts Avenue, Cambridge, MA 02139
mitpress.mit.edu

EU Authorised Representative:
Easy Access System Europe, Mustamäe tee 50,
10621 Tallinn, Estonia
gpsr.requests@easproject.com

Printed by Integrated Books International,
United States of America

HOW TO

LET

EXCUS

PA

A BIN

QUESTIONS
WITHOUT
ANSWERS

QUESTIONS WITHOUT ANSWERS

SARAH MANGUSO

ILLUSTRATED BY
LIANA FINCK

HOGARTH
NEW YORK / LONDON

PUBLISHED IN THE UNITED STATES BY HOGARTH, AN IMPRINT
OF RANDOM HOUSE, A DIVISION OF PENGUIN RANDOM HOUSE LLC,
NEW YORK.

HOGARTH IS A TRADEMARK OF THE RANDOM HOUSE GROUP
LIMITED, AND THE H COLOPHON IS A TRADEMARK OF
PENGUIN RANDOM HOUSE LLC.

HARDBACK ISBN 978-0-593-73362-2
EBOOK ISBN 978-0-593-73363-9

PRINTED IN THE UNITED STATES OF AMERICA ON
ACID-FREE PAPER

RANDOMHOUSEBOOKS.COM

2 4 6 8 9 7 5 3 1

FIRST EDITION

FOR SAM
— SM

FOR RAYMIE
— LF

CONTENTS

PREFACE

Popular culture sells parenthood as an experience of sweet, meek, mind-numbing duty, a selfless erasure in service to the next generation. I was prepared for big feelings, a love like none other, quiet music swelling as I nestled the beloved babe.

I wasn't at all prepared for how intellectually interesting it was to spend time with a young child. I realized happily that it was a project in translation, in how to understand and communicate with a creature who possessed intelligence but limited experience. The baby had a particular cry for milk that differed from his other cries. When he was slightly older but still preverbal, when he was finished eating, he waved his hands over his plate. He used that same signal when he wanted to stop whatever activity he was doing—playing a game, getting his teeth brushed. Sometimes I used the sign to tell him that we needed to stop doing something. It never failed to register.

I learned that children are dizzyingly fast-learning engines of art and experiment. I watched my child make sense of the world not as a simple-minded cherub but as a measuring, remembering machine. The idea that children were actually less limited than adults, smarter than we are in every measurable way except in accumulated experience, was humblingly new to me.

Before I started spending time around children, I thought that people who paid close attention to these simpletons were people who had decided not to be interesting anymore. I thought that people found their own children fascinating simply because they'd been biologically hypnotized into loving them. Once I learned what children are really like, I immediately wanted to create an artifact of their weird eloquence, which was such a surprise to me when I finally noticed it. During this period of my education, my son, Sam Chapman, was my first and most essential teacher.

Before he was capable of conversation, I thought I already knew what a preschooler would say because I'd seen it represented so tediously in advertisements and crappy entertainment. But by the time Sam was about four years old, I was writing down almost everything he said, shedding my indoctrination as I went.

He was indeed cute, too, as when he tried to pick up a freckle from my forearm or played at feeding a piece of pancake to his toy truck, but the cute things he did always had a hint of the abyss about them. His most interesting questions all seemed, in some way, to be about death; it was Leanne Shapton who pointed this out to me plainly.

In 2021 I opened a Twitter account and posted a single tweet: *What's the best question a kid ever asked you?* Within twenty-four hours I had more than a hundred questions. Within a week I had hundreds more. I asked some famous friends to retweet the tweet. I asked everyone I knew to ask everyone they knew. I read multiple iterations of the if-God-is-everywhere conundrum and multiple accusations of pregnant women having eaten their babies, but I also had the privilege of reading a lot of accidental poetry and philosophy in multiple languages. Death was a common topic. So was birth. I was surprised to receive so many questions about necks and chins.

The poet Kenneth Koch, who taught at Columbia University for many years, also taught poetry to young children at New York City's P.S. 61 in the late 1960s. He wrote three books on teaching poetry to young children, including many examples of the poems that arose from his prompts. When I teach creative writing, I always bring in a few poems from his anthology *Wishes, Lies, and Dreams* and mix them in with my more recognizable texts. I love asking, with a straight face, if anyone has read the work of so-and-so, and then to announce that the author was seven years old when she wrote the poem we've just discussed. My students are impressed and a bit unnerved.

Very young children, younger even than Koch's elementary school students, might not be old enough to write poems themselves, but they have access to imaginative worlds that are just as interesting, and just as notable a fount of worthwhile literature.

The word *literature* might first suggest to the lay reader a Shakespeare play or a Tolstoy novel, but I've spent much of my writing life composing and delighting in very short literary forms, among them the short poem, the very short story, and the aphorism. There's always something a bit magical about one great stand-alone line. The aphorist James Richardson writes, "No one will ever write a novel by accident. A poem, too, takes time. But if I say 'Pick a word' and you say one, where did it come from? You certainly don't say you 'wrote it' or 'created it'—more like you chose it, or it chose you. One-liners must be in the middle of that spectrum." One-liners, among them the one-line questions that constitute this book, float somewhere between thinking and writing, where verbal but preliterate young children dwell. That place is the origin of this text.

My chief purpose, in assembling the text of this book, is to challenge the popular depiction of children as adorable idiots, instead portraying them as they are: intelligent, intuitive, inventive, philosophical, funny. Their questions are a work of found choral phi-losophy, a collective subjectivity that disappears from most people's lives by kindergarten.

Like all minimalist art, Liana Finck's drawings look easy because they are simple, but her work's apparent looseness and spontaneity belie the rigor from which it is borne. She's a master at rendering huge concepts with a few strokes of the pen and maybe a word or two. Her sensibility is a perfect match for this text: playful, quick, brilliant, weird. She contains big feelings in small packages, and she resists received ideas as completely as children do, though in the latter case it's only because they haven't heard of them yet.

Young children are here with us, yet they maintain an easy ac-cess to the supernatural realm—they walk in both worlds. Most preschoolers don't yet understand that they're performing for an adult audience that desperately wants to write them off as senti-mental. The questions I included in this book are the ones that seemed the least contaminated by that adult audience. These questions are cute by the word's original definition, swift and pierc-ing. They cut to the quick.

I return to these questions when I need a little effortless wis-dom. Their ease with the abyss comforts me. They present corpses, rocks, beards, and graves as more or less emotionally equivalent, and they show me that anything can be interesting if you look at it from the right angle. They remind me that when I feel bound up and inarticulate, when I have nothing to add, I too might begin with a question.

Sarah Manguso
Los Angeles, 2023

QUESTIONS
WITHOUT
ANSWERS

PEOPLE

HOW DO YOU KNOW WHEN IT'S TIME TO LEAVE
A PARTY?

WHEN DO WE GET TO BE BABIES AND START OVER AGAIN?

DO BURGLARS LIVE IN TIRES? THE HOLES IN TIRES?

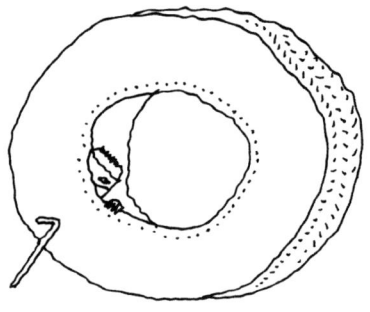

WHAT DOES A PRESIDENT EAT?

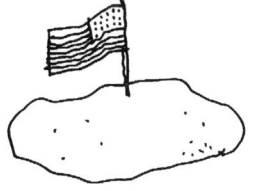

WHY DO PEOPLE LIKE MUSIC?

WHAT WERE HUMANS BRED TO DO? WHAT ARE WE
GOOD FOR?

HOW IMPORTANT ARE NECKS?

WHEN YOU DIE, WHAT DO YOUR EYES DO?

HOW DOES DATING WORK? DO ADULTS JUST
CLAIM EACH OTHER?

IF NEIGHBORS MOVE AWAY, ARE YOU STILL NEIGHBORS?

WHY ARE THE WAITERS WEARING BIRTHDAY-PARTY SUITS?

DO CLOWNS PEE BLUE?

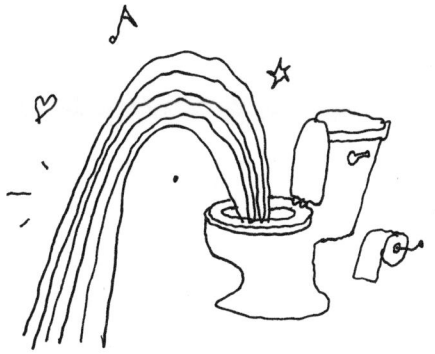

ARE WE RICH?

WHEN THE BABY IS BORN, HOW DO THE PARENTS KNOW ITS NAME?

WHAT IF YOU HAVE NO ONE TO TEACH YOU TO DRIVE?

WHAT MAKES SOMEONE UGLY?

WAS MOM A BABY ONCE, TOO? DID I PLAY WITH HER?

WILL I GROW TALL LIKE PAPA? WILL I GROW A BEARD
LIKE HIM? WILL WE HAVE TO FIGHT TO SEE WHICH
ONE OF US IS THE REAL PAPA?

DID CAVEMEN CLEAN THEIR CAVES?

WHY ARE THERE JUST BONES IN THE GRAVEYARD?
WHO TOOK THEIR SKIN?

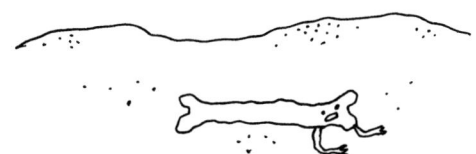

CAN A PERSON DIE OF SADNESS?

HOW DO YOU GROW UP?

WHAT MAKES A LANGUAGE OURS?

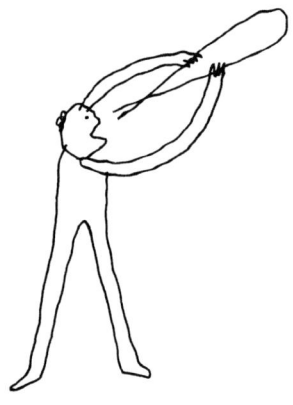

WHAT SHOULD YOU DO ON THE LAST DAY OF
YOUR LIFE?

WHY DID PEOPLE STOP BELIEVING IN THE GREEK GODS?

WHY DOES A GHOST WANDER?

ANIMALS

IS THAT DINOSAUR DEAD?

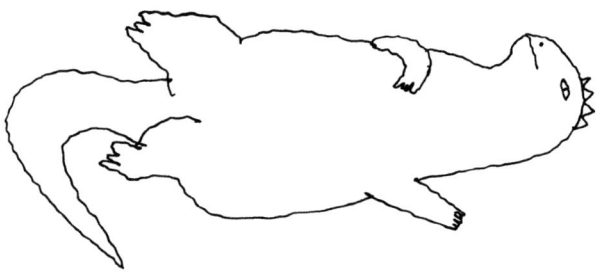

IF THE KITTY SLEEPS IN MY BED, WHY ISN'T
SHE IN MY DREAMS?

WHY DO BUTTERFLIES STINK?

WHAT'S INSIDE A CAT?

HOW DO YOU GET THE MEAT OFF THE ANIMAL
WITHOUT HURTING IT?

DO CATERPILLARS KNOW THEY'RE GOING TO BE
BUTTERFLIES, OR DO THEY BUILD THE COCOON
NOT KNOWING WHAT WILL HAPPEN?

ARE CATS RELIGIOUS?

WHAT IS PURPOSE?

CAN I TRY ON YOUR MOLE?

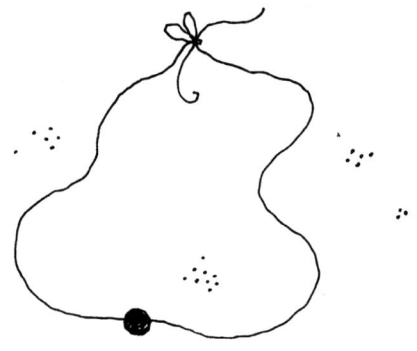

WHY DID ANIMAL CONTROL SEND A PERSON
TO HELP THAT SQUIRREL? WASN'T THERE
AN ANIMAL THAT COULD HELP HIM?

WHEN THE KITTY DIED, WHY DIDN'T WE
ALL DIE, TOO?

DO YOU REMEMBER WHEN WE SAW REAL
DINOSAURS IN MY DREAM?

DO DOGS HAVE CHINS?

IS IT DARK INSIDE AN EGG?

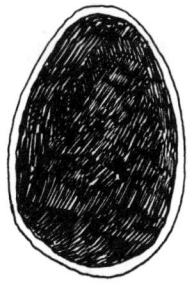

IS THE CAT MY BROTHER OR MY BABY?

DO ANIMALS TALK ABOUT US THE WAY WE TALK
ABOUT THEM?

DO YOU THINK BIRDS GET SCARED WHEN THEY GO UP
TOO HIGH?

DID THE DINOSAURS LOVE ONE ANOTHER?

DID HORSES KNOW THEY WERE IN A WAR?

DO YOU THINK OUR LIVES ARE ALL IN AN ANIMAL'S
DREAM?

WHAT'S THE OPPOSITE OF A WHALE?

THINGS

WHAT DOES A TREE MEAN?

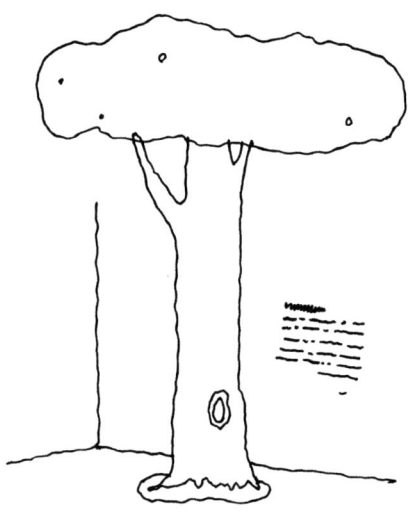

DOES A GHOST REMEMBER HOW IT FEELS
TO HAVE SKIN?

IF I PUT A BOOGER UNDER MY PILLOW,

WILL IT GROW?

ARE BUBBLES IN DRINKS THEIR THOUGHTS?

HOW DO YOU THINK SINKS WORK?

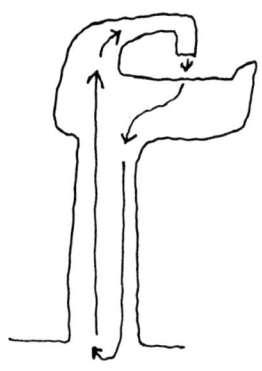

DO DIFFERENT COLOR INKS HAVE DIFFERENT
TASTES?

IS MOUNT EVEREST THE WORLD'S NOSE?

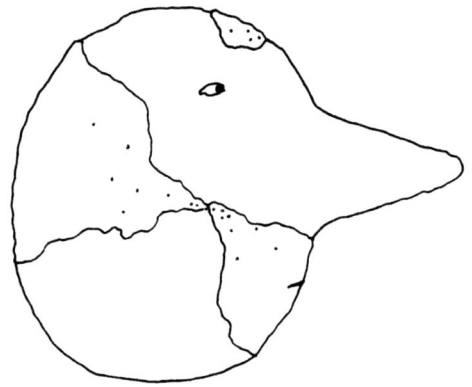

WHY IS MY JACKET TRYING TO BLOW AWAY?

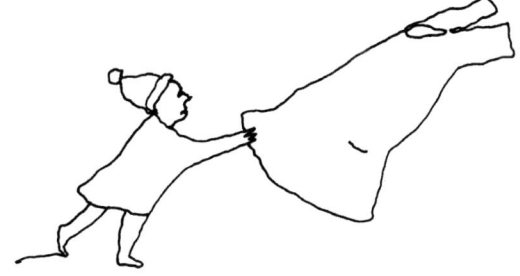

WHAT WOULD HAPPEN IF YOU LEFT A DOLL ON A CHAIR IN A HOUSE FOREVER?

DOES THE FLOOR LIKE TO BE STEPPED ON?

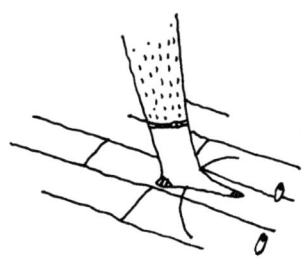

DOES THE SUN HAVE A MOTHER?

IS THERE A SPOT IN OUR HOUSE THAT NO ONE
HAS EVER WALKED ON?

CAN I EAT A HEART?

ARE BEARDS GOOD?

IF EVERYTHING BECOMES A GHOST, WOULDN'T THERE BE SO MANY GHOSTS THAT WE'D BE SUFFOCATED BY GHOSTS?

IS A NIGHTLIGHT THERE TO TELL YOU HOW LONG
THE NIGHT IS?

DO PEOPLE BURY OLD TOYS?

WHAT DOES A GARGOYLE SAY?

WHAT DO CLOUDS TASTE LIKE?

WHAT IS A COUNTRY?

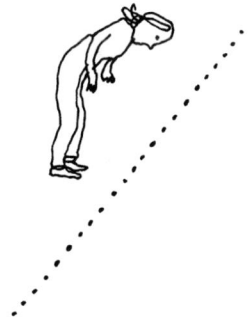

DO YOU LIKE WINDOWS?

WHAT IS IT LIKE INSIDE A LIGHTBULB?

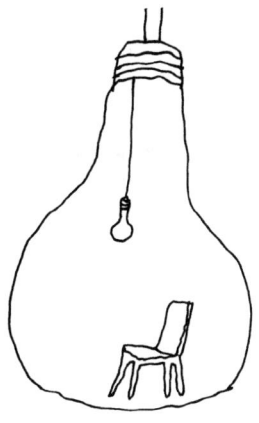

DO YOU HAVE TO SLEEP WITH YOUR CRUTCHES?

WHERE DID IT COME FROM? (WHERE DID
WHAT COME FROM?) MY BODY.

BEFORE EARTH, WHAT WAS THERE?

CAN IT RAIN ON THE MOON?

DO CLOCKS GET DEPRESSED AT NIGHT?

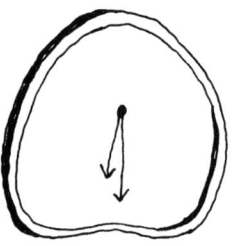

WHAT IS THE DARKEST THING IN THE WORLD?

WHERE IS THE INTERNET?

BIG THINGS

IS O THE OPPOSITE OF 1 BECAUSE IT DOESN'T
HAVE ANY STRAIGHT LINES?

WHAT IS A MOMENT?

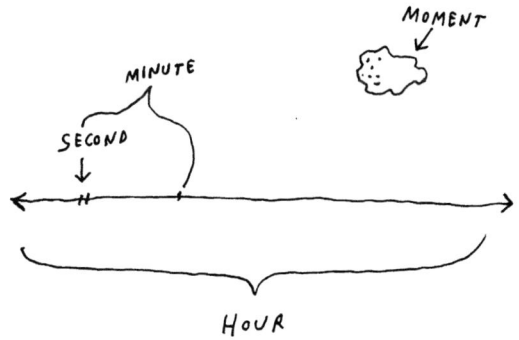

MOMENT

MINUTE

SECOND

HOUR

WHAT IS POINT A?

POINT A

CAN SOME THINGS NOT BE DESCRIBED?

WHAT PUSHES THE RIVER?

IF YOU OPEN THE SUNROOF OF THE CAR AT NIGHT, WON'T ALL OF THE DARKNESS COME IN?

WHAT'S THE DIFFERENCE BETWEEN STOP
AND WAIT?

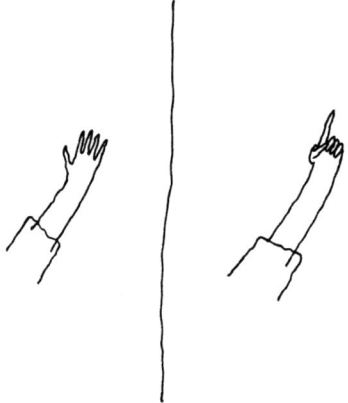

WHAT IS BUSINESS?

IS THE WIND JUST THE SKY COMING TO YOU?

WHAT'S A DARK IDEA?

WHAT WAS THE FIRST SONG?

WHY IS EVERY DAY THE SAME BUT DIFFERENT?

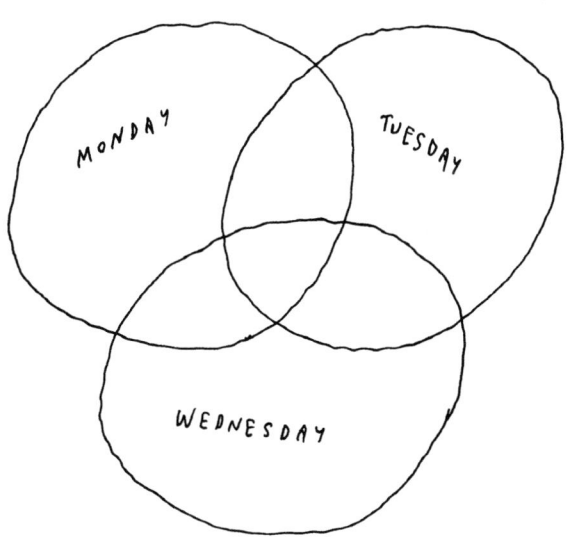

IF BIG ROCKS BREAK INTO SMALL ROCKS AND SMALL ROCKS BREAK INTO PEBBLES, WHY HAVEN'T WE RUN OUT OF BIG ROCKS BY NOW? IS THE WORLD JUST NOT VERY OLD?

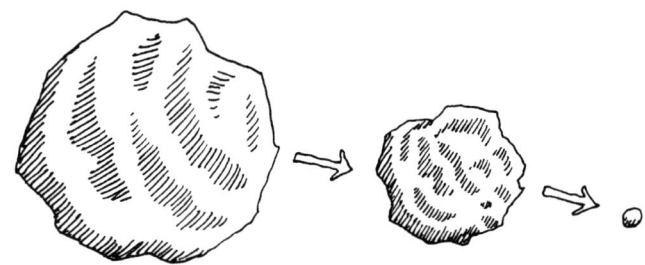

DOES THINKING COUNT AS WORK?

IS THE WORLD TRYING TO START ITSELF OVER?

DO YOU THINK ALL MOVIES ARE MADE UP?

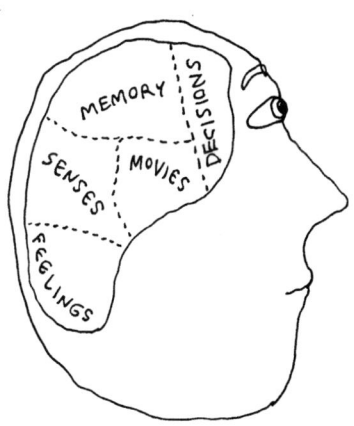

HOW DO PEOPLE GET MADE? FROM OTHER PEOPLE, OR FROM THE FOG?

WHAT'S THAT TEAR IN THE SKY?

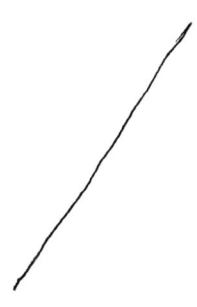

IS THE AIR GROWING?

WHAT IS HISTORY?

IS TIME JUST LIKE LITTLE BLOCKS IN SPACE?

WHAT IS MORE IMPORTANT, LOVE OR HOPE?

WHERE DOES THE DARK GO WHEN THE
LIGHT COMES ON?

WHICH SEASON COMES FIRST?

WHY DO THE DAYS KEEP COMING?

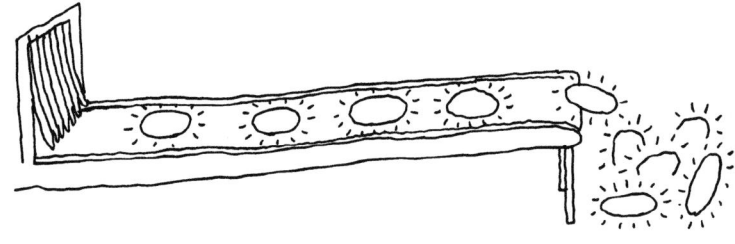

WERE YOU BORN WITH A MUSTACHE?

WHY ARE YOU SO TALL?

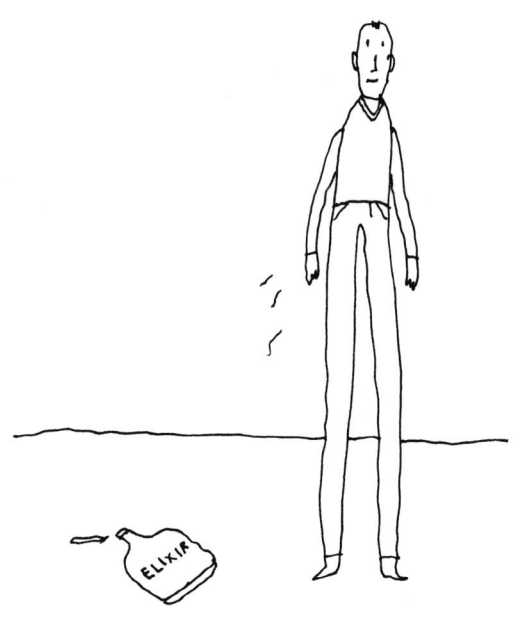

WHY DON'T PARENTS JUST PLAY SPORTS
THEMSELVES?

WHEN I'M BIG, WILL YOU BE THE LITTLE MOTHER?

WHICH LEG DID THE BABY COME OUT OF?

WHEN I SAW A SHADOW AND CHASED IT DOWN THE HALL, WAS IT YOU?

WHY DO YOU HAVE A ROCK IN YOUR HOUSE?

WHAT IF YOU LEAVE ONE DAY?

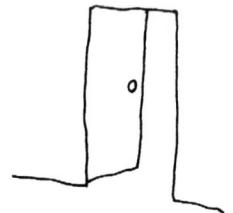

WHEN I WAS IN YOUR BODY, DID YOU KNOW ME? WAS
I EXCITED TO MEET YOU?

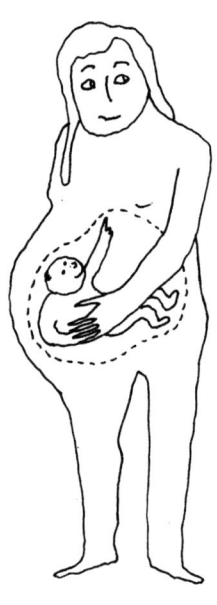

DID I MAKE THE WORLD?

IF I DIDN'T NEED YOU ANYMORE, WOULD YOU DIE?

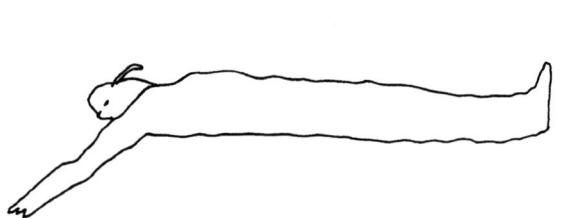

HOW OLD WERE YOU WHEN YOU WERE BORN?

WHY IS THAT THE HUSBAND THAT YOU CHOSE?

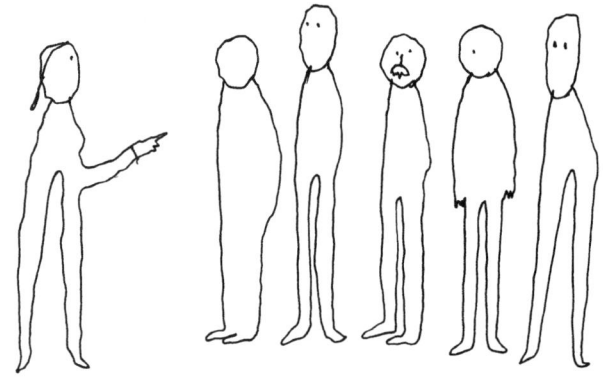

ARE YOU REALLY MY DAD, OR ARE YOU JUST SOME
GUY WHO HELPS ME PUT ON MY PAJAMAS AND STUFF?

WHAT WAS THE FIRST TIME YOU WERE A
GROWN-UP?

ARE YOU THE SAME PERSON WHEN YOU'RE SLEEPING?

WHEN YOU DIE, CAN I COME WITH YOU?

AFTER THEY BURY YOU, WHEN DO THEY COME BACK
AND DIG YOU OUT AGAIN?

HOW OLD ARE YOU WHEN YOUR HANDS SWITCH PLACES?

WHEN DID YOU LEARN TO DANCE?

WHAT WOULD IT TAKE FOR YOU TO BE HAPPY?

ARE YOU HAPPY?

DO YOU HAVE A DARK SECRET?

DO YOU LIKE BEING OLD?

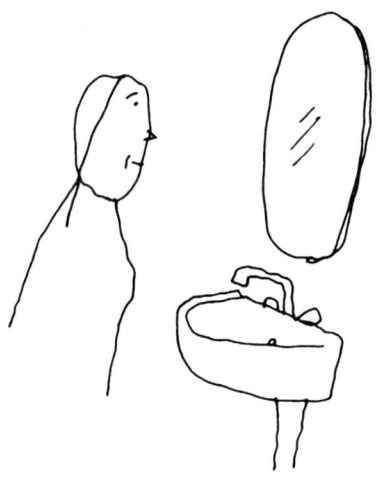

WHAT'S THE WORST THING YOU'VE EVER DONE?

ME

CAN YOU HEAR MY THOUGHTS?

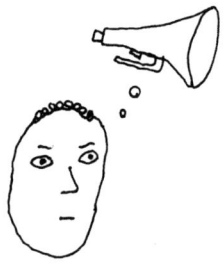

HOW MUCH MUD CAN I EAT?

SHOULD I TRY NOT TO READ PEOPLE'S MINDS?

IF I SUCK MY FINGERS FOR LONG ENOUGH, WILL MY NAILS GET POINTY?

WHEN I GROW UP, WILL I STILL HAVE THE
SAME NAME?

WHY CAN'T I CHANGE MY BIRTHDAY?

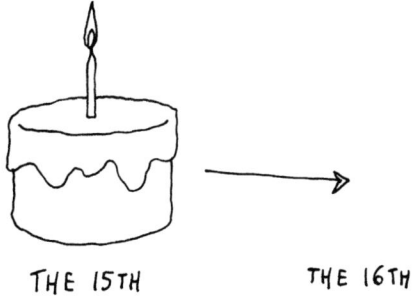

THE 15TH   THE 16TH

AM I ADOPTED? HOW DO YOU KNOW?

WHEN I WAS BORN, DID YOU CHOOSE ME OR DID
I CHOOSE YOU?

I ALWAYS THOUGHT WHEN I DIED I WOULD BE IN
A PIECE OF ICE. IS THAT TRUE?

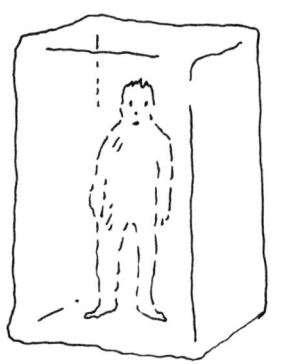

WHY AM I CRYING?

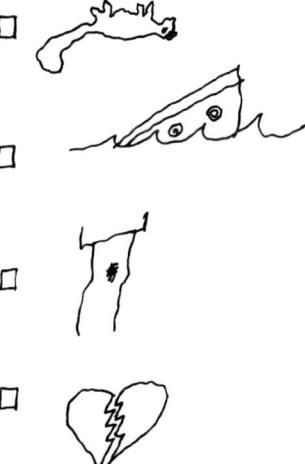

WHEN I DIE DO I JUST LIE UNDER THE DIRT
UNDER THE GROUND?

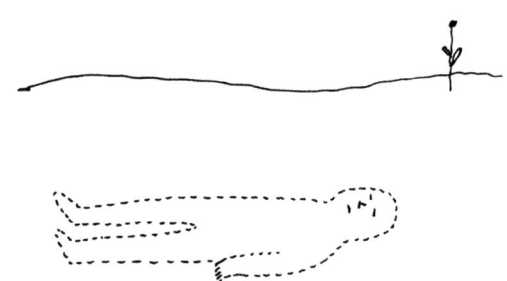

HOW DO I KNOW IF I'M HAPPY?

WHEN I WAS REALLY LITTLE, WHERE DID YOU
FIND ME?

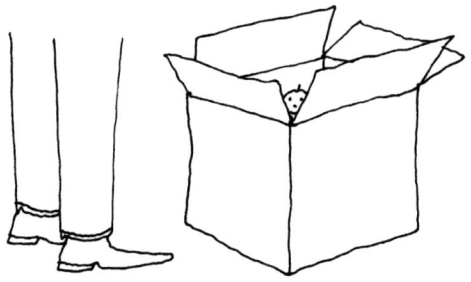

IF YOU CUT OFF MY HEAD REALLY QUICKLY,
WOULD I STILL KNOW I WAS DEAD?

HOW WILL IT FEEL ON THE LAST DAY
I'M A CHILD?

TWINGE

DO I HAVE TO BE LYING DOWN TO DIE?

HOW CAN YOU BE YOUR OWN FRIEND?

HOW DID I GET INSIDE MY BODY?
WHAT HAPPENS TO THE PEOPLE WHEN I WAKE UP?
WHY DO YOU WANT ANOTHER BABY?
WHAT'S THE PASSWORD TO THE MICROWAVE?
IS GRANDMA A SKELETON YET?
IF YOU COULD ONLY SAY ONE WORD FOR THE REST OF YOUR LIFE, WHAT WOULD YOU SAY?
WHAT HAPPENS AT THE END OF THE STORY OF THIS LIFE?

I WISH TO THANK MY AGENT, PJ MARK; MY EDITORS, PARISA EBRAHIMI AND ANDY WARD; AND MY INTREPID RESEARCH ASSISTANTS, GISELLE FAHIMIAN CHANDLER AND MEGAN WILLIAMS, ABOVE ALL OTHERS. I ALSO ACKNOWLEDGE THE GENEROUS SUPPORT OF SAM CHAPMAN, ROB CORDDRY, ELIZABETH DOAN, ANDREW SEAN GREER, DANIEL HANDLER, SHEILA HETI, MIRANDA JULY, JENNIFER L. KNOX, MAILE MELOY, B.J. NOVAK, JULIE ORRINGER, LEANNE SHAPTON, AND ANTOINE WILSON.

— SM

A GIANT THANK-YOU TO ANDY WARD AND PARISA EBRAHIMI; MY AGENT, MEREDITH KAFFEL SIMONOFF; AND CASSIE GONZALES, SARAH FEIGHTNER, ERIN RICHARDS, AND THE REST OF OUR TEAM AT HOGARTH AND RANDOM HOUSE.

— LF

SARAH MANGUSO IS THE AUTHOR OF NINE BOOKS, MOST RECENTLY THE NOVELS *LIARS* AND *VERY COLD PEOPLE*. HER OTHER BOOKS INCLUDE A STORY COLLECTION, TWO POETRY COLLECTIONS, AND SEVERAL ACCLAIMED WORKS OF NONFICTION. HER WORK HAS BEEN RECOGNIZED BY AN AMERICAN ACADEMY OF ARTS AND LETTERS LITERATURE AWARD, A GUGGENHEIM FELLOWSHIP, A HODDER FELLOWSHIP, AND THE ROME PRIZE. SHE LIVES IN LOS ANGELES.

LIANA FINCK IS A CARTOONIST LIVING IN BROOKLYN. SHE IS THE AUTHOR OF *LET THERE BE LIGHT, PASSING FOR HUMAN*, AND *EXCUSE ME*; A CHILDREN'S BOOK, *YOU BROKE IT*; AND A MEMOIR ABOUT MOTHERHOOD, *HOW TO BABY*. SHE IS A REGULAR CONTRIBUTOR TO *THE NEW YORKER*. SHE IS A RECIPIENT OF A GUGGENHEIM FELLOWSHIP, A FULBRIGHT FELLOWSHIP, AND A WALLANT AWARD.